A BLUEPRINT FOR CREATING A TOXIN-FREE LIVING ENVIRONMENT

healthier homes

JEN & RUSTY STOUT

VICTORY BELT PUBLISHING INC.

LAS VEGAS

Cover illustration by W. R. Barrineau & Associates Architects

Cover design by Kat Lannom

Interior design and illustrations by Yordan Terziev and Boryana Yordanova

Photography by Robert G. Gomez Photography and Tyler Schmitt, Mill Photo Studio

Printed in Canada

TC 0122

To my wife, Jen:

While my name may be next to yours on the cover of this book, I unabashedly give you all the credit for making this dream a reality.

We always say the homes we build change lives. Today, without hesitation, I can say that no event has changed me more for the better than the day I fell in love with you! Through all the obstacles and health problems thrown your way, you made your own path with grace and changed an industry along the way.

You have always been an inspiration to me, and I am beyond excited to share that inspiration with the world.

I love you more than bacon.

–Rusty

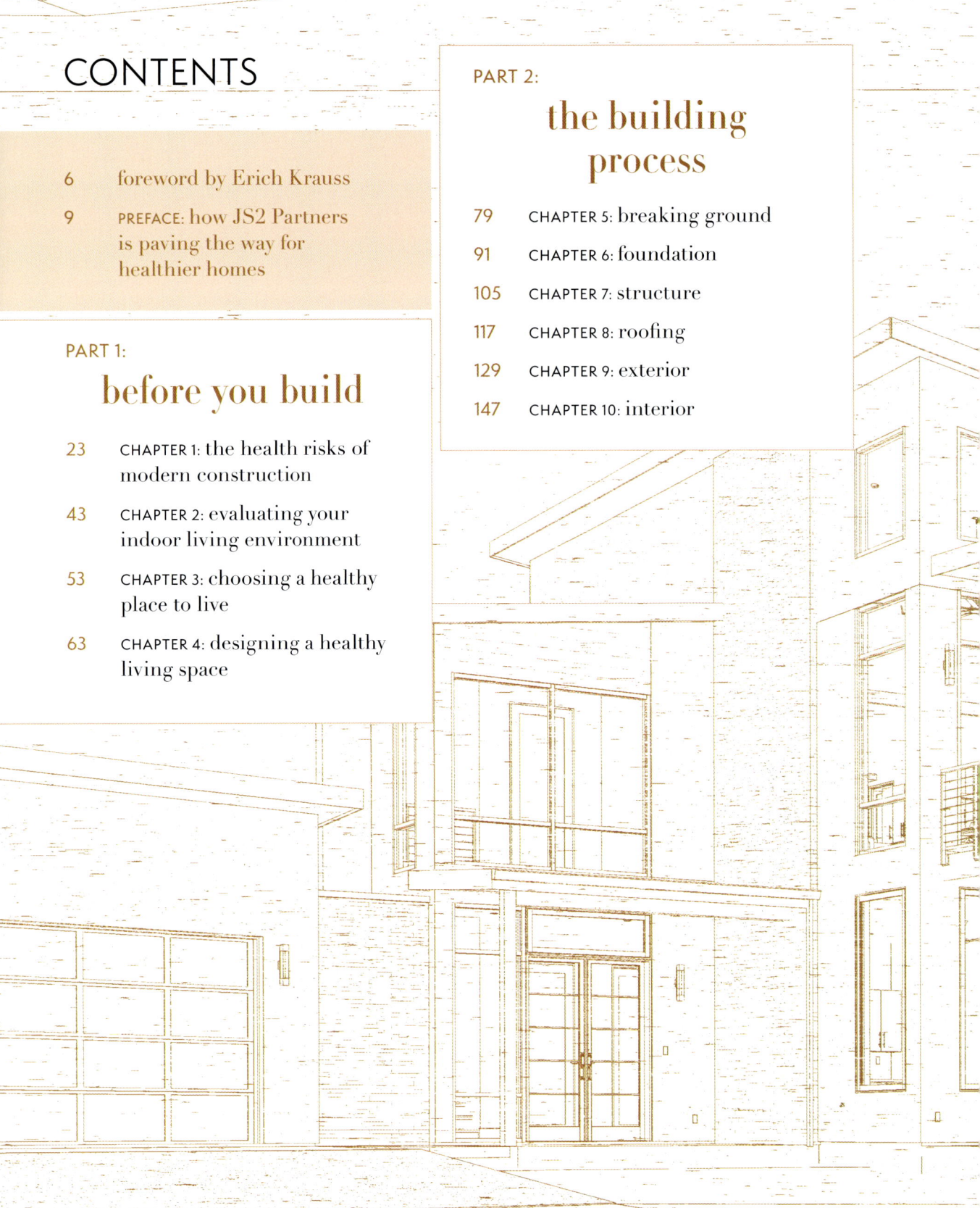

CONTENTS

FOREWORD

Ten years ago, if someone told me that a house could make you physically sick, I would have smiled, said, "Sure, sure," and then walked away quickly. After all, I had spent most of my life in remote parts of the world—the Amazon rain forest, the jungles of Thailand, and the plains of Africa. These places were filled with microbes that evolved to feed on flesh, and yet I had always returned unscathed. I figured a healthy diet and clean living could ward off the worst the natural world had to dish out.

Turns out, I was wrong.

Over the course of a few years, my wife and I started experiencing strange symptoms—anxiety, muscle pain, blurred vision. We both figured it was stress and carried on. Then, one day, while I was working out in our home gym, something broke. Setting down the weights after just a few reps, I desperately needed to take a nap. I awoke in what I can only describe as a different body. Every muscle and joint hurt, it felt like an icepick had been embedded in the back of my neck, and the anxiety I had been carrying turned into crippling panic. Little did I know, chronic illness would define our lives for the better part of a decade, all thanks to a moldy patch of drywall beneath a sink.

Mold illness, otherwise known as chronic inflammatory response syndrome (CIRS), put my immune system into a tailspin, activating latent Lyme disease I must have been carrying since my youth. Suddenly I was sensitive to just about everything—food and exercise and toxins, many of which were coming from my own home in the form of volatile organic compounds (VOCs). We knew if we were ever going to heal, we needed a clean living environment, so we decided to build a home from scratch using only nontoxic materials. It seemed like a simple enough goal. We couldn't have been more wrong.

There were very few books on the subject available at the time, and those that were available read like rocket science manuals. I thought, *even if I learn everything these manuals have to offer, what builder is going to take the time to get up to speed?* There had to be a better way, so I started reaching out to healthy building consultants who could lead my builder and me through the entire process. I was absolutely thrilled until our first group meeting. The consultant began rambling on, sharing the chemical compositions of different materials but not supplying any practical applications. My builder's eyelids were half-closed at the end of the meeting, and he hadn't absorbed a thing. So, I paid the consultant's fee and tried another. Same thing.

My builder was ready to ditch me as a client, and truthfully, I was about ready to give up on the dream of building a home that would support my health rather than harm it. In a last-ditch effort, I stumbled upon a husband-and-wife team out of Texas and decided to give it one more try. It was the best decision I ever made.

During our first meeting, it was already apparent Jen and Rusty weren't trying to impress us with how much they knew about building healthy homes. They knew that most builders' bandwidths for trying new things were small or nonexistent. So, they cut to the chase, letting us all know which materials were toxic, which nontoxic materials they could be replaced with, and how those products needed to be installed. They made the whole process so simple and pain-free, even my builder got excited about what we were creating.

When the building process started, Rusty and Jen were part of each weekly builder's meeting, and construction went off without a hitch. Radon mitigation went in first, then the nontoxic foundation, then the mold-free lumber got hammered together to form the frame. Then the real fun began—mold-proof sheetrock, nontoxic paint, and no-VOC cabinets. The whole process became a thrill ride, and soon my wife and I were scooping up all the information we could get. We installed healthcare facility wiring to avoid EMFs, whole-home air exchangers, and water shutoff switches on every outlet to detect a leak and automatically shut off the water. Truthfully, the whole healthy-home project became an addiction, and when all was said and done, we even had a freshwater pool.

Now I consider myself a healthy-home expert, all thanks to Rusty and Jen. They made me realize that you don't have to be a chemist to learn what will make you sick. Their approach was so simple, literally anyone could follow their guidance. It seemed to me that this was something the world needed, and so I approached them with the idea of writing this book. And now, dear reader, you have their expert guidance at your fingertips. Those early manuals I attempted to read contributed to a good amount of hair loss—either from the stress or from tugging it out from the roots—but the book currently in your hands will be an entertaining ride, guaranteed. You'll learn how to design, build, remodel, and decorate a house that will support your family's health in the years to come, and I hope you find that as empowering as we did.

Erich Krauss,
President, Victory Belt Publishing

preface

how JS2 Partners is paving the way for healthier homes

Jen Stout, co-founder of JS2 Partners and Healthier Homes, discusses why modern homes are toxic.

Have you ever woken up to a bad hair day? You get out of bed, look in the mirror, and realize that your locks are all pasted to one side of your head, while the other side looks as though you stuck your finger in an electrical socket. We can all relate to bad hair days.

But have you ever woken up to no hair at all? Not a single strand on your head—not even eyebrows or eyelashes? Well, it is a terrifying feeling. And that's what I woke up to for years in my twenties. As a young, physically fit lady with plenty of vitality, tenacity, and enthusiasm for life, I never thought I could become so broken.

I was twenty-six years old and living in Dallas, Texas, in a swanky new uptown apartment. As an ambitious young woman, I decided to resign from my job as the marketing and events manager at one of the nation's largest pediatric hospitals and pursue my dream of going to graduate school to earn my master's in business administration (MBA).

I started working part time as a freelance marketing consultant while studying my tail off for the Graduate Management Admission Test and applying to MBA programs. As a former journalism major with a marketing background, I almost couldn't believe it when I was accepted to Southern Methodist University's graduate program at the Cox School of Business.

Working from home in my Dallas apartment was an adjustment, but my setup was great. I had a little desk in a nice sunny spot with my computer, books, organized file folders, and schedule. But, not long after the transition to spending most of my time at home, I noticed some drastic changes in my health.

First, it became harder to concentrate. I chalked it up to being distracted in my own space. But then my sleep schedule started creeping later and later into the night. Some nights, I couldn't fall asleep until 4 a.m., which led to getting up later as well. My face began swelling up for no obvious reason, and my eyes and throat burned. I got rashes and hives on my face, arms, and legs. I'd suddenly become allergic to everything.

I got rid of my soaps and makeup, switching to more natural products. I remember going to Sephora at North Park Mall with no makeup on, the skin on my face red with irritation, and feeling so embarrassed as I asked which brands were best for sensitive skin.

After that, foods I'd enjoyed my whole life started making my mouth swell up. I had no history of food allergies except nuts, but suddenly I was forced to give up more and more favorites, such as pineapple, chicken, limes, black pepper, and mint.

The rashes were getting worse, and my hair was falling out. I had to wear pants and long-sleeved shirts and disguise my hair loss with creative hairdos and scarfs. After I developed skin allergies to cotton and polyester, my wardrobe slowly converted to alternative fabrics like bamboo and Tencel, which were easier to tolerate.

I tried to pretend nothing was wrong and keep my head in the game. I had made it into SMU's MBA program, and nothing was going to stop me from excelling in grad school. This was my one shot.

Meanwhile, I saw infectious disease doctors, dermatologists, rheumatologists, ENTs, and even specialists in other states. No one had a clue what was causing my symptoms. They put me on lupus medication even though my lab results showed I had no autoimmune disorders.

My hair eventually all fell out, so I finally sucked it up, drove to North Dallas, and bought a wig. One of the older ladies in the store asked if I was going through chemo. I politely shook my head no, with tears in my eyes. What had my life come to? And why could no medical professional help me?

Shortly before my lease ran out on the uptown apartment, I found a clinic in Dallas that specialized in environmental medicine. I had never heard of environmental medicine or chemical sensitivities before, but I was desperate, and its reputation was good; people from all over the world would spend months there for treatment.

The doctors immediately suspected mold toxicity. Sure enough, my lab tests came back positive for high levels of mycotoxins, which are toxic secondary metabolites produced by fungus that can cause disease and even death in humans and animals. I had dangerous levels of trichothecenes and ochratoxins, two mycotoxins that are particularly harmful. I was the proverbial canary in the coal mine after my exposure to these vapors.

I started searching my apartment and found black-colored mold behind the baseboards around my shower. Upon removing the baseboards, I realized the sheetrock and walls were damp to the touch—moisture had wicked up the walls surrounding the entire shower. On the opposite side of that wall was my desk, where I spent most of my days and nights studying and working.

 GEEK BOX | different types of mold

Not all mold is harmful to humans. Some drugs such as antibiotics were developed from molds, and molds give certain cheeses their distinct flavors. Mold also plays a vital environmental role in breaking down decaying organic material.

That said, if you start developing symptoms such as fever, night sweats, weight loss, rashes, year-round allergies, and/or respiratory difficulties that doctors can't seem to figure out, it might be time to investigate your living space.

S. chartarum is a particularly nasty species that's more commonly known as "toxic black mold." This is the same type of mold that was linked to the highly publicized Cleveland case in the 1990s, where several babies developed bleeding of the lungs and died.

I did some tape-lift samples, and the apartment building manager independently hired a mold remediation specialist to perform testing. When my samples came back positive, I called the remediation specialist. He confirmed that he too found several types of mold, not only in the bathroom but also along the back wall of my closet. Apparently, my shower pan wasn't the only one leaking, and the closet's carpet was saturated from a similar leak in the neighboring apartment. During our phone call, the mold specialist said, "Between you and me, I would get out—now. You have Stachybotrys chartarum. That's the deadly kind."

When it became clear that the pans were defective and cracked upon installation—which meant they'd been leaking for years, not just in my unit but in all the surrounding apartments—the management office relocated me to another building for the remainder of my lease. To this day, I vividly remember the clean-up crew entering my apartment dressed head to toe in Tyvek protective suits, goggles, and gas masks. All my belongings were boxed up and taken to a special facility to be professionally cleaned in giant hydroxyl chambers. It was finals week, and I wore the same dress the entire time, since all my belongings were gone, and few stores carried clothes made of bamboo or lyocell fabrics back then.

Despite the toll that mold exposure took on her health, Jen graduated from Southern Methodist University with a master's in business administration.

Meanwhile, I proudly graduated from SMU with a master's in business administration. But instead of using my hard-earned MBA to start applying for cool new jobs, I prepared to leave Dallas and move back home to Houston to be closer to friends and family. My health was failing, and I needed help and support, which was a hard pill to swallow after being so independent for so much of my life.

I spent the next few years driving with my mom back to Dallas once a month to receive medical treatment and tests for allergy shots. During that time, I learned that my immune system had completely crashed, I had become allergic to everything around me, and it was going to be a long road to recovery. My body had gone haywire due to the years of exposure to toxic mold. After being constantly bombarded for so long, it had started reacting to everything around me.

what are mycotoxins, and why are they dangerous?

Scientists have found countless fungal (mold) species throughout the world that produce a spectacular array of toxins to counteract predators and minimize competition from other organisms.[1] Mycotoxins are the specific class of fungal toxins that affect humans and animals.

Many are bioaerosols, meaning they spread through vapor. Several are potent carcinogens (for example, aflatoxin from A. flavus and Aspergillus parasiticus), while others include trichothecenes produced by the Fusarium and Stachybotrys species, fumonisins and zearalenone produced by the Fusarium species, and ochratoxin A produced by Aspergillus ochraceus and Penicillium verrucosum. Although many mycotoxins are immunotoxic, the trichothecene mycotoxins are also immuno-stimulating at lower doses.[2]

My central nervous system was overwhelmed, and my body was no longer able to detoxify itself. This led to a buildup of chemicals, heavy metals, and volatile organic compounds (VOCs) in my tissues. My bloodwork was full of dangerous toxins, such as ethylbenzene, xylene, cadmium, lead, and organo-phosphate pesticides. My immune system markers were all out of whack, and my inflammation markers were spiraling out of control.

I was a young person with no previous medical issues, so everyone was baffled when I simultaneously developed thyroid issues, anemia, and vascular problems. I even managed to contract Lyme disease from tick bites not once but twice during this time. Most healthy immune systems can fight off Lyme disease, but mine was far from healthy.

Throughout my years of detox, I learned from heavy metal tests and chelation provocations that I was also carrying around a toxic load of lead, which had accumulated in my bones because of exposure to lead dust in an improperly ventilated attic conversion in my childhood home. My dad was a shooting instructor and would make his shotgun shells from scratch on an assembly carousel, which explained the gray dust covering all our toys and the ever-present aroma of gun oil and "pencils." Surprising to think that two decades of lead exposure as a child didn't do nearly as much damage as just a few years of exposure to mycotoxins as an adult.

Early on my road to recovery, I realized that I needed a place to live where my body could heal—a place where I wouldn't have allergic reactions to the smoke from the restaurant next door, the chemicals from the glues used to put down the engineered wood flooring, and the fumes from the paint on the walls and ceiling.

Jen's toxic mold exposure caused imbalances in her oxygen levels, immune system, and nutrition uptake. She wore a wig and a PICC line for two years.

I needed a place where my canary body could halt its fight-or-flight response and enter a state of rest and relaxation. The problem was that such a place did not exist.

For example, the slightest amount of formaldehyde in the air would set off a cascade of inflammation and rashes all over my body—and practically all modern buildings contain materials laden with formaldehyde. In newer homes or apartment buildings, the materials would still be "off-gassing," which means emitting harmful gases as a by-product of industrial processing. In older homes, the slightest amount of mold or mildew in the HVAC system or behind the walls would set my body off. I needed to figure out a way to build a new home that was free from chemical toxins and mold. So I put my journalism school research skills to good use.

I began researching and designing my first home in 2012. It was a long learning process and a first-of-its-kind healthy home. My builder was a veteran at crafting custom homes, and he and his son worked in lockstep with me throughout the process. It was a new journey for them as well, but they were just as dedicated to the end goal as I was.

I researched every nut, bolt, and screw. I looked at thousands of safety data sheets and probably put in enough hours to earn an honorary degree in chemistry along the way. I contacted countless manufacturers of lumber, concrete, mortar, adhesives, insulation, drywall, tape, duct work, appliances, plywood, roofing, underlayments, sheathing materials, paints and stains, bathtubs, tile, grout—the list goes on and on.

I sat down with the architect and put careful thought into the home's layout and proximity to neighbors. I learned the processes at each step and familiarized myself with every material and its purpose in the construction process. I tested new construction methods and considered a number of alternative building materials. That said, we were careful to build the home using only proven, high-quality, structurally sound materials.

Some materials were more costly than standard items used in high-quality custom homes—but others were actually less expensive. Some items had labels such as GREENGUARD Certification, which means it meets certain chemical emissions standards, while others didn't. I couldn't afford to rely on product certifications. If a single product with off-gassing potential slipped through, the whole home would be rendered unhealthy and intolerable for me.

WHAT IS A SAFETY DATA SHEET (SDS)?

An SDS is a standardized document mandated by the International Hazard Communication Standard, which requires chemical manufacturers to compile and communicate data related to occupational safety and health. These documents typically contain information such as health and environmental hazards and safety precautions for dealing with chemicals.

We were careful to wash off any materials that could've been contaminated during the manufacturing process, such as the machined metalwork for the ducts. We inspected all items before installation and caught some big issues before they became permanent, such as the wrong drywall package being delivered.

After tons of brainstorming, we learned how to ask the important questions. For example, what are the best options for whole-home air-purification systems that do not use heavy adhesives in their filters but will allow you to run the fresh air intake through the filter before the air enters the home? For my home, we decided to place these intakes on the side away from the neighbors' dryer exhaust vent. And rather than putting the intakes on the shingled roof, where a hot summer day could warm the asphalt and create tar fumes, we decided to pull from under the eaves. These were different ways of doing construction and very important considerations! I couldn't believe no one else was paying attention to these details.

Finishing the home was bittersweet. I had accomplished the impossible: a completely clean and nontoxic house that wouldn't make my ever-so-sensitive body react. But, upon completion of my first healthy home, I had to make the difficult decision to sell the house. Between the Gulf Coast region's high humidity, Houston's air pollution, and the elevated levels of electromagnetic radiation (also called "electrosmog"), I decided it was best for me to move out of the big city.

I decided to move to Horseshoe Bay, a small resort town about forty-five minutes from Austin. I had always loved the central Texas Hill Country region, with its lakes, hills, and fresh country air. My family spent every summer there when I was growing up, so it was a very special place to me.

I moved into my family's house on one of the many lakes in the area. The home had its fair share of issues (even after going through an extensive mold remediation process several years prior), but it was my best option until I could build my next healthy home, and for that I was thankful.

At the time, I had three part-time gigs: I was running my own freelance marketing and PR firm. I was also working as a healthy home consultant on various builds. And I spent the rest of my time healing through a mix of holistic medicine, like infrared sauna sessions and supplementation, while also using conventional treatments like antibiotics and allergy shots. It was a regimented long-term program, and I was slowly starting to reverse the damage, grow my hair back, and regain a portion of my energy. I feel incredibly blessed that I had access to the resources that helped me heal.

When I was finally able to return to full-time work, I got a job as the executive director at the Hill Country Builders Association (HCBA). With my new income stream to back me up, I gathered my savings, found a lot with a lake view, and set out to build my own healthy home—again. But this time I was going to get to live in it!

I met my husband, Rusty, a veteran builder of twenty-plus years, through the builder's association. He was president of the HCBA board of directors when I started with the association, and he soon became integral to designing and building what is now our healthy home. He learned how to build healthy from me, and I learned how to be a builder from him.

Through me, Rusty witnessed firsthand the struggles that can result from living in a moldy or otherwise unhealthy environment. He saw not only what I had gone through but also what a tremendous difference it made in my energy, allergies, and quality of life when we moved into our brand-new healthy home. Knowing that there are thousands out there just like me, not to mention millions who just want to live healthier lives, Rusty and I decided to give others the opportunity to have their own healthy home. Thus, our company JS2 Partners Healthy Home Builder was born!

A lot has happened since then. For starters, we had our first baby—who we lovingly refer to as the #babymonster. Rusty and I have worked with dozens of homeowners, built many healthy homes from the ground up, and taken on several challenging renovations. And now we want to bring our knowledge to you—to give you the tools you need to have the right conversations with your builder so that you can create the healthy home you deserve. That's why we started Healthier Homes.

The time has come for all of us to take a closer look at our indoor environments. We spend most of our time indoors, and the quality of the air we breathe, the water we drink and bathe in, and the materials that surround us have a profound impact on our bodily functions, our brains' acuity, our ability to be creative and productive, and our quality of life.

The toxins in our environments affect everyone on varying levels, and just because you're healthy now doesn't mean you'll never struggle with the effects of the harmful materials around you. In fact, over 50 percent of Americans have one or more gene variants that can cause them to be genetically predisposed to processing toxins ineffectively.[3,4] There are several important enzymes involved in metabolizing nutrients and breaking down harmful VOCs such as mycotoxins from mold. The PON1 and MTHFR genetic variants are among the polymorphisms that have been found to affect these important enzyme-driven detoxification processes.

Jen met Rusty through the Hill Country Builders Association (HCBA)—they're seen here accepting an award for the HCBA's youth building trades program at the Texas Association of Builders banquet.

Chronic illnesses and cancers that used to be rare are becoming exponentially more common, likely driven by our environment, our genetics, and our growing toxic burden. Now more than ever before, homes are sealed up and made airtight according to strict energy code requirements, which has created concentrated environments for off-gassing. Even if everything in your home is GREENGUARD Certified, chemicals like formaldehyde are still allowed in these products—just at lower levels. Imagine the long-term impact on human and animal health with constant VOC emissions from flooring, furniture, bedding, paint, adhesives, insulation, drywall, etc., circulating inside your home. It all adds up quickly.

Many of us pay a premium for organic and non-GMO foods, electric vehicles, and gym memberships. Why don't we want to pay a premium for a healthy living environment? Perhaps many are simply unaware that their current living environment is suboptimal. Whatever the reason, the longer you wait to make a change, the less viable that excuse becomes.

GREEN VERSUS HEALTHY

Remember, "green" does not mean "healthy." Many recycled or repurposed materials have been fumigated or contain pesticide and petroleum residues. Many recycled plastic products will forever smell like fragranced laundry detergent. While we fully support sustainability and eco-conscious efforts, our number one priority is occupant health.

Rusty and I founded JS2 Partners on a mission to create better living through healthy building. Now, we've created Healthier Homes as an invitation for you to join us as we explore construction best practices and guide you toward health and prosperity for you and your family—now and into the future.

You don't need to be a builder to understand healthy building, and you don't need to be an interior designer to source healthy furnishings. Whether you're already aware that your home is causing health issues or you're among the fortunate individuals who have yet to feel the adverse effects of an unhealthy living environment, this book is for you. Inside, we will break down the design and build process into bite-sized steps presented in the simplest and most forthright way possible. We will share multiple options whenever possible, pointing out the pros and cons of each. When you're done, you will know how to converse with your builder to get what you want out of the building process, and you'll know how to furnish your home safely and beautifully.

And that process really does start with you. You're the homeowner, the occupant, the parent, the pet owner. It's up to you to demand that homes, schools, and workplaces be built with healthy materials and methods that promote quality of life and well-being for our fellow humans and pet friends.

The choice is yours! So, Rusty and I invite you to join us on our journey to building a healthier future, together, for everyone.

Jen & Rusty Stout

Healthy Home Builders

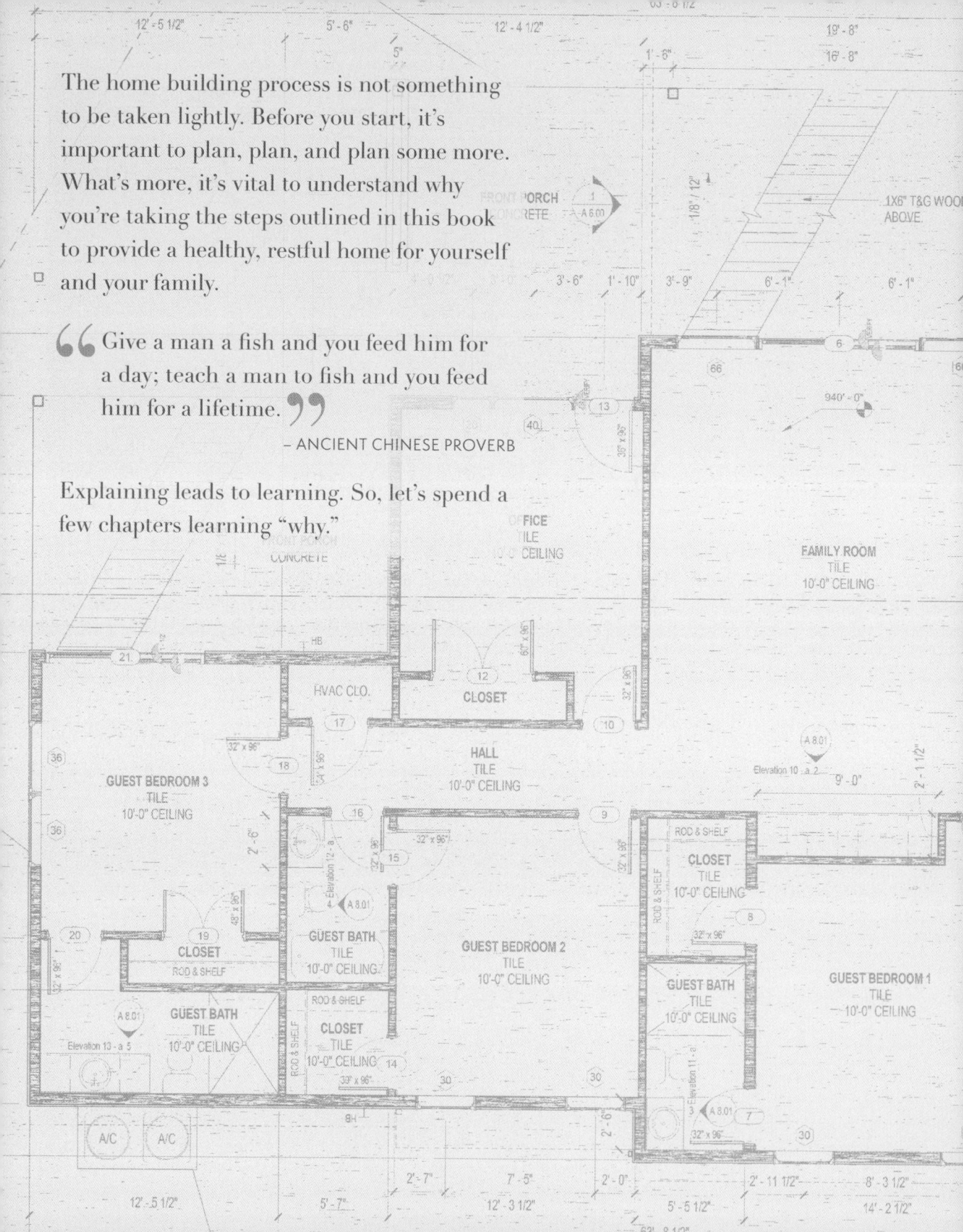

The home building process is not something to be taken lightly. Before you start, it's important to plan, plan, and plan some more. What's more, it's vital to understand why you're taking the steps outlined in this book to provide a healthy, restful home for yourself and your family.

> " Give a man a fish and you feed him for a day; teach a man to fish and you feed him for a lifetime. "
>
> – ANCIENT CHINESE PROVERB

Explaining leads to learning. So, let's spend a few chapters learning "why."

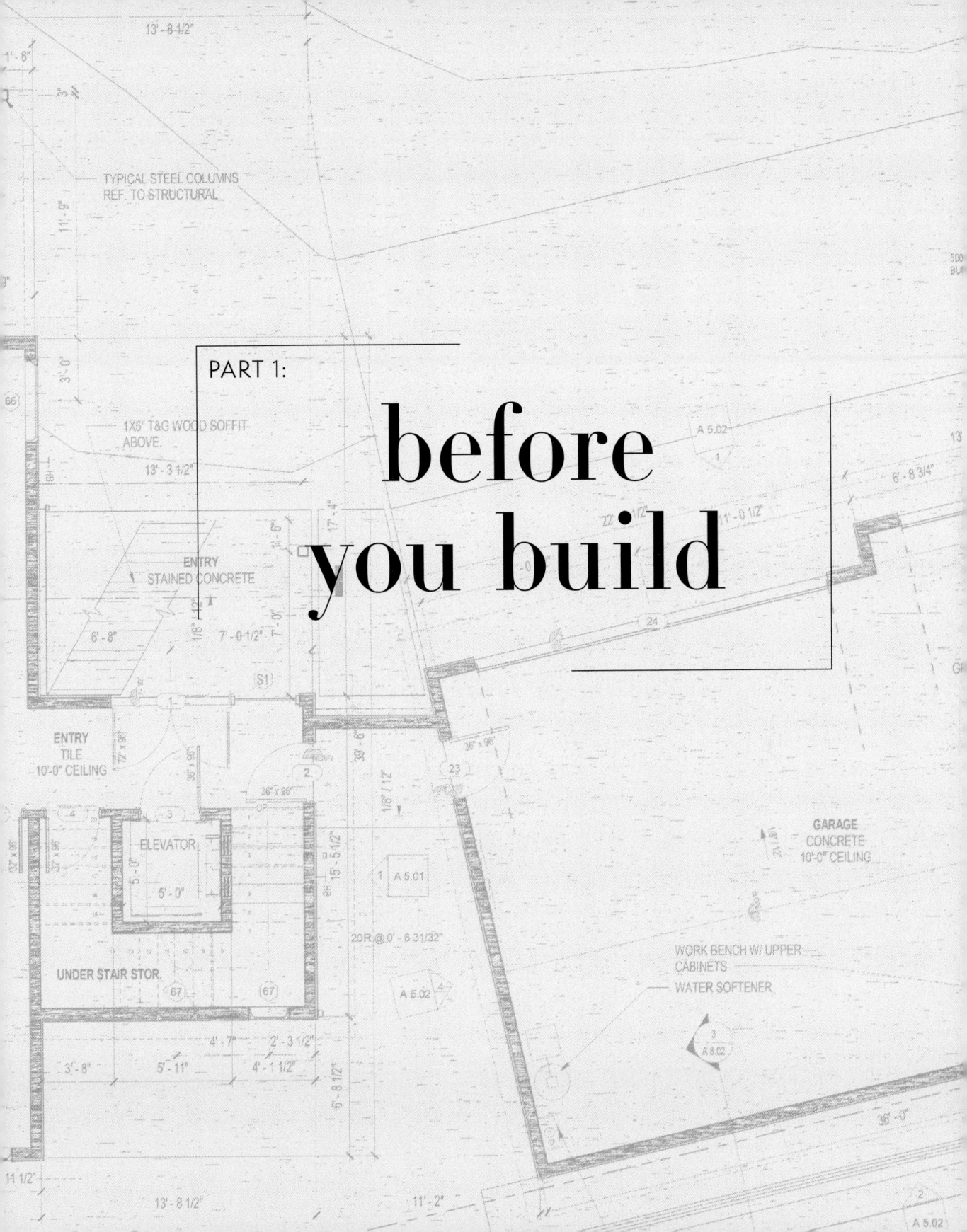

PART 1:

before
you build

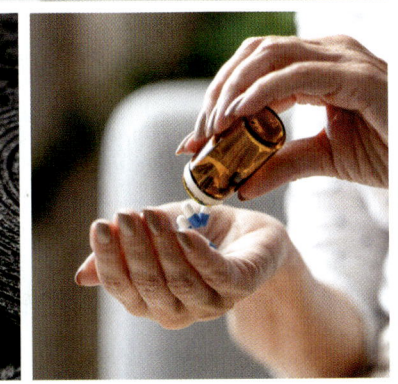

the health risks of modern construction

WHY WE BUILD HEALTHY

We are healthy home builders, but what does that mean? "Healthy" can be a relative term, like how too much salt may be harmful for one person while another may benefit from additional sodium in their diet.

The World Health Organization (WHO) defines "healthy" as a state of complete physical, mental, and social well-being, and not merely the absence of disease.[1] This is the fundamental concept we apply to our building approach. So, let's break things down based on our primary goal: to build living environments that promote wellness for the occupants.

"Wellness"—now there's a word we are passionate about. Wellness means being in good health, a state of being that most healthy people actively pursue as a lifelong goal. So, we're actually wellness builders. We create better living through healthy building by putting each homeowner's health and wellness front and center.

That's why we wrote *Healthier Homes* as an easy-to-read guide on how to build a healthy and safe indoor living environment. We want this book to be straightforward and informative for everyone, whether you're new to the world of construction, interior design, and healthy living or you're a seasoned expert. As we walk through the home-building process, we will present concepts, ideas, and solutions to everyday problems in the simplest terms possible. By the end, you'll know how to work with your builder to make your indoor living space more conducive to well-being and productivity, and you'll have the knowledge to make informed decisions when furnishing your home.

We are passionate about this topic because we live it. Our mission was born out of necessity. When Jen's health started failing while she was studying at SMU, she eventually realized that an indoor living environment with toxic black mold and chemical off-gassing from cheap building materials was to blame. Her own home was literally poisoning her. Bottom line, our homes should not be making us sick!

what is off-gassing?

During manufacturing, products such as furniture and building materials are made with raw materials that often contain chemicals called volatile organic compounds (VOCs), which are slowly released into the air through "off-gassing."

In humans, exposure to VOCs may cause negative health effects such as

· Allergic reactions

· Headaches

· Respiratory problems

· Hormone disruptions

· Serious illnesses such as cancer

Off-gassing levels are higher in new construction and new home furnishings. Higher temperatures and humidity levels can accelerate the off-gassing process, meaning VOCs can accumulate faster during the hot summer months, especially inside enclosed, airtight spaces like modern houses.

As a start, combat these problems by

· Replacing off-gassing items with low- and no-VOC alternatives

· Using air purification systems to help filter out some VOCs

· Increasing ventilation by using fans and opening windows

· Keeping humidity in check with a whole-home dehumidification system

· Sourcing materials free of what the US Environmental Protection Agency deems HAPs (hazardous air pollutants)

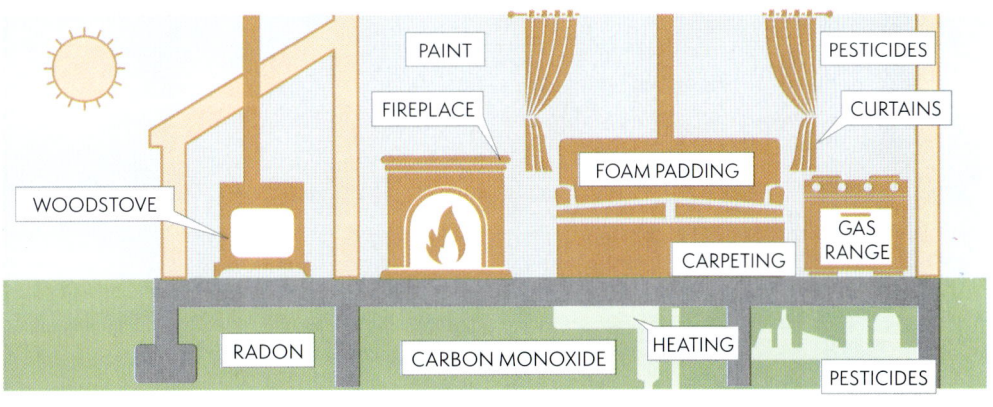

GETTING EDUCATED

With so many building materials on the market, it can be hard to know what's behind the walls of your home. Plus, modern manufacturing methods have enabled a lack of visibility into the many harmful substances present in most building products. So, when we set out to do something no one else was doing—build healthier homes using exclusively nontoxic construction materials and methods—we started by educating ourselves. Knowledge is power, after all.

HORSESHOE BAY **FEBRUARY 21, 2019**

PRSRV STD
U.S. POSTAGE
PAID
MARBLE FALLS, TX
78654
PERMIT NO. 50

BEACON

VOL. 13, ISSUE 4 • PO BOX 4845 • HORSESHOE BAY, TX 78657 • thebeacon@nctv.com

HSB Duo Pioneering Healthy Home Building

By Jodi Lehman

Home builders Rusty and Jen Stout are doings things differently with a mission to pioneer home building best practices that put people's health first. The Horseshoe Bay husband and wife team recently launched JS2 Partners, LLC to craft homes using chemical free materials, low and no VOC components, and integrated building science designs that promote health and well-being for their homeowner clients across the U.S.

"Americans spend on average 90% of their time living and working indoors, and recently the EPA found indoor air to be twice as polluted as outdoor air. Many are unaware of the harmful health effects of chemicals found in nearly all modernday construction materials used in homes, schools and the workplace," said Jen Stout, JS2 Partners cofounder. "I'm living proof that moving into a less toxic living environment can drastically improve one's health."

Jen was a victim to indoor toxicity years ago when her health rapidly declined while living in Dallas and pursuing her MBA from Southern Methodist University. Jen's sudden onset of a failing immune system, food allergies and sensitivities to clothing and cleaning products left doctors puzzled until they found mold residues in Jen's

Home builders Jen and Rusty Stout envision health and productivity as the essential building blocks for crafting homes that are functional, safe and beautiful for their clients. Their business, JS2 Partners, is located in the Center Point Building at 6909 W. FM 2147 in Horseshoe Bay.

medical lab tests. This pivotal finding soon led to the discovery of extensive black mold growth behind the walls of her apartment. Jen knew she had to get into a clean and healthy environment, which is when she embarked on her journey into researching and building

healthy homes.

"Our priority is to provide clients a place to call home that's conducive to health and productivity. These two factors are the essential building blocks for crafting homes that are functional, safe, comfortable and beautiful," said

Rusty Stout, long-time builder and co-founder of JS2 Partners. "Our clients are nationwide, ranging from families with allergies and sensitivities to health-minded individuals seeking a new custom

See Home
Continued on Page 9

Because the public lacks reliable information when it comes to modern construction and the health implications of our living environments, a part of our mission as healthy home builders is to raise awareness so that people can live better, healthier lives. The information we share in this book will help you make informed decisions about your well-being and quality of life within your own home.

When it comes to understanding the fundamental link between your living environment and your health, it all boils down to how the human body works. It's important to have at least a basic understanding of how humans and animals process substances from our environments so that you can begin to pinpoint areas of concern around you. So, here's an ultra-fast lesson in biology.

The central nervous system (your brain and spinal cord) communicates with the peripheral nervous system (your organs, limbs, skin, muscles, and so on) via thousands of electrical pulses every second. This internal communication system works via sympathetic and parasympathetic responses, which are often referred to as the state of *fight or flight* and the state of *relaxation,* respectively.

Any form of stress will keep our bodies in fight-or-flight mode. Stress can come from many things—strained relationships, pressure at work, or external stressors caused by our living environments, for example. Often, that third category goes unrecognized.

Harmful environmental stressors include VOC emissions and mold vapors in the air we breathe, heavy metals in the water we drink and use for bathing, and wireless microwaves emitted from "smart home" technologies. These pollutants interfere with the fundamental ways our bodies are supposed to function.

sympathetic versus parasympathetic responses

The human body has an autonomic nervous system (ANS) that controls its involuntary functions, such as heart rate, breathing, and blood flow. Within the ANS are two subsystems:

· The parasympathetic nervous system controls "rest and digest" functions.

· The sympathetic nervous system is responsible for "fight or flight" functions.

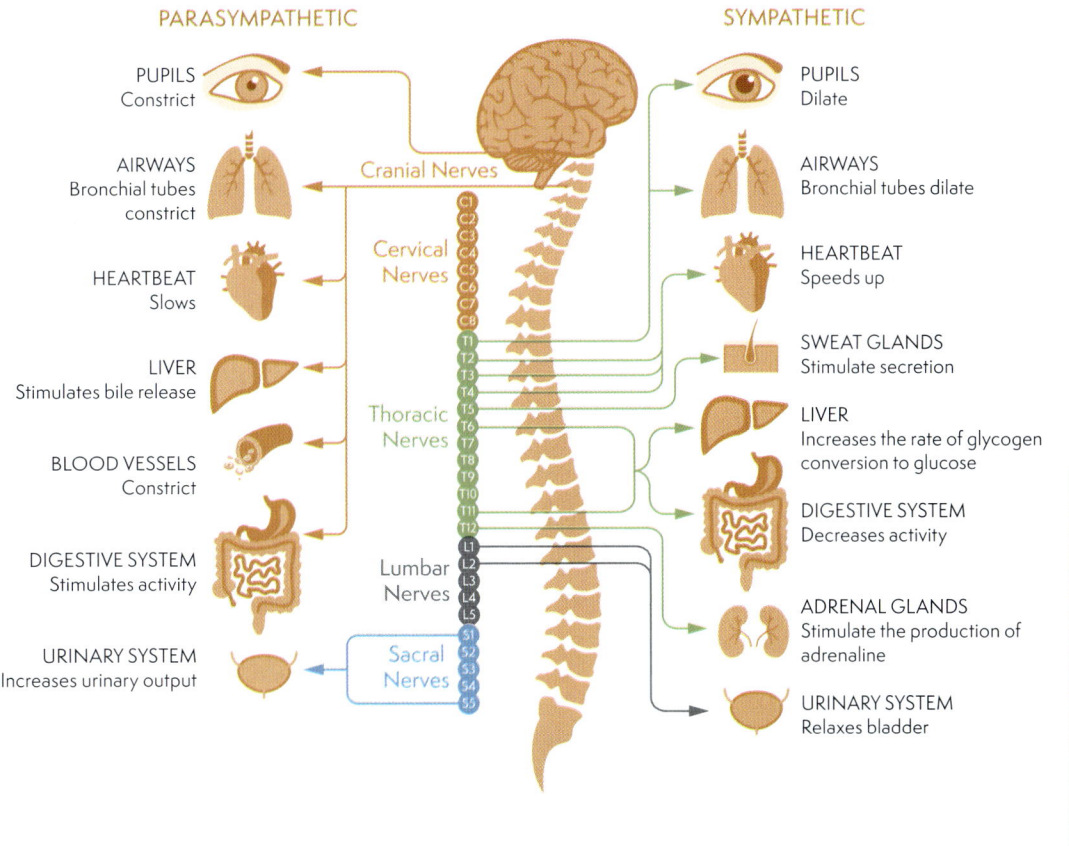

On a cellular level, a body stuck in stress mode will stay in the sympathetic fight-or-flight state. In that state, the body loses the ability to rest well and detoxify itself efficiently, leading to sleep deprivation and eventual energy production problems. Our bodies begin to literally break down, just as a car would break down without routine oil changes, tire rotations, and engine tune-ups.

It's no wonder that the rate of chronic illness and cancer has continued to rise at an alarming pace over the past several decades.

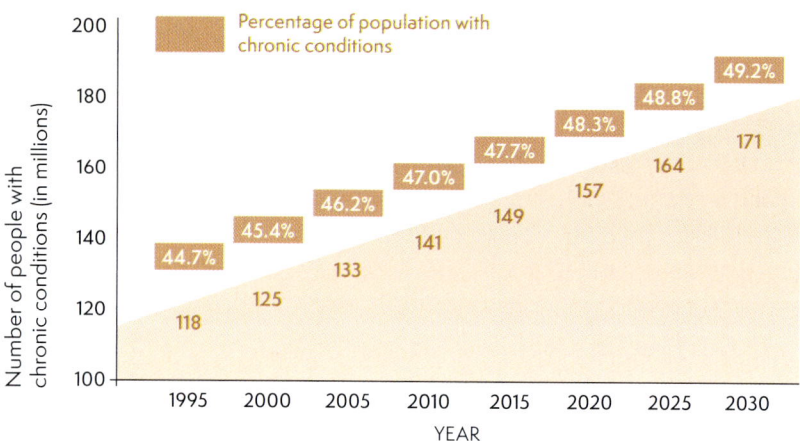

Source: Shin-Yi Wu and, Anthony Green, "Projection of Chronic Illness Prevalence and Cost Inflation," RAND Corporation, October 2000.

Inevitably, stress happens in our lives. We work twelve-hour shifts, stay up all night with a newborn, or pull all-nighters before final exams. We fight rush hour traffic and inhale the exhaust fumes from cars and trucks. That's life, and it's OK—but our homes should be places that promote rest, rejuvenation, and productivity. To stay healthy, we need somewhere to decompress, sleep, and function optimally, which can only happen in the parasympathetic state.

HOW MODERN CONSTRUCTION COMPROMISES OUR HEALTH

A variety of modern construction practices create stress on our bodies. For one, stricter energy codes call for homes to be built airtight. This means the VOCs emitted by construction materials are continuously compounding to exponentially pollute the indoor air.

The 2019 *Homes for Health Report* from the Harvard T. H. Chan School of Public Health states that concentrations of air pollutants are two to five times higher indoors compared to outside.[2] These VOC emissions are directly linked to a laundry list of chronic health issues, including cancers, cardiovascular disease, and respiratory conditions. At a global level, the WHO states that seven of the ten leading causes of deaths in 2019 were due to noncommunicable diseases. The top three causes were ischemic heart disease, stroke, and chronic obstructive pulmonary disease.[3]

What does this have to do with housing? Vinyl chloride, toluene, benzene, formaldehyde, chloroform, hydrazine, naphthalene, acrylonitrile petroleum hydrocarbons, and dichloroethane are all found in numerous building products. These dangerous chemicals are known for off-gassing, and the Agency for Toxic Substances and Disease Registry (ATSDR) includes them in its list of substances that produce toxic effects on human cardiovascular, respiratory, and hematological systems.

7 Out of 10 Deaths Globally Due to Noncommunicable Disease

When you put two and two together, there's no denying that chronic VOC exposure has a profound impact on our health.

Airtight construction also causes water vapors from showering, cooking, and breathing to continually build up inside air-conditioned spaces. Humid air creates the perfect breeding ground for mold and mildew inside a home, which is the culprit behind another class of VOCs called mycotoxins.

As we discussed earlier, mycotoxins are secondary metabolites produced by filamentous fungi, aka the mold found in homes with moisture issues. Mycotoxin vapors can be absorbed via skin contact and through inhalation of spore-borne toxins in the air. Mycotoxin emissions from mold are toxic to the liver, central nervous system, renal glands, and reproductive and immune systems.

So, you may be wondering who is going to do something about this epic problem? In the US, some people assume that the Environmental Protection Agency (EPA) will step in to help. However, the EPA was founded in part to police *outdoor* air and water quality. Currently, there are no federally mandated indoor air quality standards. So, as long as outdoor emissions from factories are kept to acceptable levels, manufacturers can use basically anything they want in construction materials.

Well, not *anything*. In the past forty years, the EPA has banned a total of four new toxic chemicals. That brings the overall total to nine chemicals, and asbestos isn't even one of them![4] To be fair, with more than 80,000 industrial chemicals in use today, the task of evaluating each one is overwhelming. But this shows how ineffective the current environmental watchdog system is.

Third-party organizations such as GREENGUARD and Cradle to Cradle have developed their own criteria for achieving reduced levels of chemicals in consumer products. While these certifications are wonderful, we don't believe they are stringent enough. Formaldehyde, creosote, petroleum, and polyvinyl chloride (PVC) are just a few of the many ingredients that are still allowed at varying levels according to these organizations' criteria.

Our Healthier Homes standard has zero tolerance for building with harmful and/or irritating chemicals that are released into the air, and we do our best to keep this from happening!

Here's a list of the some of the most common chemicals found in building products and their health impacts on people and pets:

Formaldehyde is used in nearly everything that goes into a home because it drastically cuts down on the curing times for paint, adhesives, molded products, and so on. For mass manufacturing, this means less production time, more material output, and larger profits. Formaldehyde also off-gasses for years and is directly linked to allergy-induced skin rashes, chronic inflammation, decreased ability to detox (due to glutathione depletion), and increased rates in cancer-related mortality.[5]

Organic solvents are used in paints, varnishes, lacquers, adhesives, cleaning agents, inks, and some polymers. Don't let the word "organic" fool you. These solvents are merely carbon-based, and many are known carcinogens and neurotoxins that poison the reproductive system.[6]

Chlorinated solvents are often found in plasticizers that give certain polymer-based materials extra flexibility. Polyvinyl chloride (PVC) is used to make faux leathers, foams, caulks, luxury vinyl flooring, and epoxies. Chlorinated solvents continually off-gas and have been found to cause dermatitis, central nervous system depression, cardiac arrhythmias, and narcosis.[7]

Petroleum distillates and synthetic rubber compounds are used throughout homes in tar-based products, adhesives, waterproofing membranes, and underlayments. Petroleum products off-gas for a very long time. Among the thousands of petrochemicals on the market, benzene, ethylbenzene, xylene, and toluene have been found to cause damage to the lymphatic, hematological, reproductive, and immune systems. These chemicals are carcinogenic via inhalation.[8]

Bisphenol A (BPA) is used in epoxy resins, which means it is found in roofing materials, paints, caulks, flooring, and fiberglass as a binder. BPA is a widely used chemical in plastic that interferes with our hormonal systems. The concern regarding its use in baby products and water bottles has gained quite a bit of press in recent years. A recent review of its toxicological effects shows that BPA can decrease fertility, impact the mammary glands, alter brain function, and cause metabolic changes that affect blood glucose and insulin levels, even at low levels of exposure.[9]

Public outcry over BPA in consumer goods prompted manufacturers of baby products to swap BPA with its close chemical cousins, BPS and BPF. Sadly, these chemicals are equally harmful, but manufacturers are more than happy to tout their products as "BPA Free." Dr. Joseph Allen, the Healthy Buildings program director at the Harvard Chan School of Public Health, calls this a "regrettable substitution."[10]

Highly fluorinated chemicals are nearly as ubiquitous as water and are found in a number of construction finishes, including stain repellents. Clothing, rain gear, dental floss, cosmetics, nonstick cookware, furniture, carpet, curtains, and paints contain these harmful "forever" chemicals (so-called because they are resistant to environmental degradation). Ever wonder why your nonstick pan loses its nonstick coating over time? Or why stain-repellent surface treatments need to be reapplied after several months? Where do you think these chemicals are going? The answer is inside you—98 percent of Americans have these chemicals in their blood.[11]

You may have seen the movie *Dark Waters*, based on the true story of an entire community poisoned by DuPont's use of highly fluorinated chemicals in the manufacturing of consumer goods. The highly publicized DuPont case was about C8, the cancer-causing nonstick chemical that was found to interfere with body weight regulation (now known as an "obesogen") and elicit "the most dramatic immune suppression ever observed for an environmental toxicant."[12,13] Think about that the next time you sit on your stain-resistant couch or let your baby play on the stain-resistant carpet.

Flame retardants are another group of bad actors. Polybrominated diphenyl ethers (PBDEs) were widely used from the 1980s to the early 2000s and were found in everything from kids' pajamas to mattresses, sofa cushions, upholstery, and television sets. These fire-retardant chemicals are found in virtually everyone today. PBDE is mistaken for T4 (a thyroid hormone) once it enters the body, interfering with thyroid function, neurological development, and reproduction.[14,15,16]

Chlorinated and Brominated Tris are two other widely used chemicals that produced serious health effects, and more recently, organophosphate fire retardants have been introduced in building materials. A regrettable substitution? You bet. Organophosphates are linked to severe adverse reproductive issues and stillbirths.[17]

The massive overuse of fire retardants was a result of Big Tobacco lobbying for flame retardants to be used in consumer goods to shift the focus away from cigarettes being the leading cause of house fires in the 1980s. Their push to limit regulations that favored self-extinguishing cigarettes was a success, as told in the *Chicago Tribune*'s 2012 "Playing with Fire" series. Have any of these dangerous chemicals been banned by federal agencies? Even with all the scientific evidence we've cited, the answer is no.

Wireless technologies in our homes have gone from sporadically used radio frequency (RF) signals to higher, more intense microwave frequencies. Old-school TV remote controls would transmit a quick RF signal to the receiver on the television to change the channel, just like a car key that would unlock the car door with a click. Now, the ability to control devices wirelessly has taken over our lives. Wi-Fi and Bluetooth capabilities can be found in everything from refrigerators, cooktops, and home security systems to televisions, beds, light bulbs, and garage doors. Our cars can talk to our dishwashers from several states away. Although this is futuristic and perhaps convenient, is it really necessary?

The Federal Communications Commission (FCC) regulates the frequency output that goes into home products in the US. While the FCC is concerned about public safety, it's more along the lines of homeland security than public health. That's because the FCC's public safety policies are focused on providing access to effective and reliable communication systems for first responders and US citizens. While there's no denying that our nation's communication infrastructure is vital to national security, that's about as far as the FCC goes in terms of public safety.[18]

Regardless, there's a reason your mom told you not to stand near the microwave in the kitchen when it was running. The microwave is essentially a box that uses electromagnetic frequencies to make water molecules inside food vibrate to produce heat, cooking the food from the inside out. Fortunately, there are strict standards for shielding on microwave ovens, but the same cannot be said for most other devices now connected via the next generation of wireless technology.

Whereas RF and infrared don't allow for constant connections and require line of sight to operate, Wi-Fi and Bluetooth are much more powerful and can penetrate walls and concrete. This allows for constant connectivity in all parts of the home. It also means your appliances are continuously emitting microwave frequencies.

Microwave radiation can penetrate deep within the human body. The brain and central nervous system are particularly vulnerable to the nasty effects of electromagnetic frequencies (EMFs), which have been linked to cancer, debilitating fatigue, and significant psychiatric symptoms such as memory problems and depression.[19] Since the FCC had already approved these frequencies for use inside satellites, microwave towers, and shielded "cooking boxes" in the kitchen, this regulatory entity turned a blind eye to amplified microwave EMF usage within the home. No testing was ever carried out by the FCC to determine whether this constant inundation of frequencies would harm people.

The reality is that an overwhelming number of independent studies have exposed the harmful effects of Wi-Fi, Bluetooth, and now 5G on the human body. Yet one of the FCC's top initiatives is to expand 5G technologies across the country. The commission is also planning to free up spectrum on even higher and more powerful bands than the traditional 2.5 and 5 GHz microwave bands of current Wi-Fi technologies, creating unlicensed spectrums at 5.9 GHz, 6 GHz, and above 95 GHz and stating that they're pushing to bypass state and local approval processes to make it happen.

As healthy home builders, we practice what we preach. Our own home has neither Wi-Fi nor Bluetooth; everything is wired. However, there's still Wi-Fi in public spaces, retail businesses, office buildings—even the guy standing next to you on the bus probably has a phone, watch, and earbuds that are all wirelessly connected. A study published in *Military Medical Research* found that as little as two hours a day of 2.45 GHz Wi-Fi exposure throughout the gestation period in pregnant rats had profound negative effects on the offspring's neurodevelopment and capacity to handle stress. The radiation disrupted the brain's biochemistry, causing the adverse effects to persist into adulthood.[19] Although this study was done on rats, we're sure you get the point— it's better to be proactive and safeguard against harmful radiation. This is why we build our homes without Bluetooth, Wi-Fi, or similar technologies. We discuss minimizing EMFs in more detail in the Energy chapter later in this book.

These technologies can be problematic in other ways. Connecting the HVAC system in a home via Wi-Fi, for instance, means that the air conditioning or heating may not work properly when the internet goes out. Imagine an entire "smart home"—which supposedly offers unparalleled convenience—becoming useless due to service outages or routine router maintenance. Then there are the security issues behind wireless technologies that can make your home's network vulnerable to hackers looking to steal your personal information.

You might think a healthy home would be even less convenient, but that's not the case. Everything in our new home construction, including the internet, communicates via wiring, making each healthy home reliable, functional, secure, and comfortable.

BUSTING THE MYTHS OF HEALTHY BUILDING

In this book, we debunk a variety of myths about healthy building. For example, if we had a dollar for every time someone mistakenly equated healthy building to using "green" or repurposed materials, we could afford to give away healthy homes to everyone.

Many people think that healthy building requires the use of straw bales, hay, clay, hemp, or a myriad of other earthen products, many of which have not been adequately reviewed for performance and would never pass local building codes. This misconception is especially troubling because some of the companies that manufacture alternative building materials claim they are nontoxic when in fact many contain unhealthy substances.

An example is an insulated concrete form (ICF) system that places recycled wood chips and cement aggregate on a wall of vertical rebar. These systems do not allow a moisture barrier to be built into walls but do enable cement fumes and wood terpenes to enter the home. Plus, walls full of rebar can amplify EMFs throughout the home.

Some builders that specialize in earthen homes (that is, homes constructed using natural materials such as hay, straw, and clay) refer to their practices as "healthy" and "green." While they may be sustainable, studies show that hay and straw are often contaminated with mold. Furthermore, Stachybotrys chartarum, a particularly problematic and potent type of black mold, will likely produce illness-inducing mycotoxins in humid conditions on hay and straw.[20] "Stachy," as we call it, is unique in that its mycotoxins are immune system stimulants. While other molds are known for suppressing the immune system, Stachy can cause the onset of severe allergies and hyper-responsive, deregulated immune systems in people and pets.[21]

Another example of a "healthy" construction product with hidden issues is mineral wool batt insulation, which goes behind walls. While manufacturers claim this insulation is inert, petroleum products are typically used in the binders. We avoid most petroleum-based products, as many produce harmful off-gassing.

Manufacturers may also get away with claiming their products are formaldehyde-free while including ingredients that act as formaldehyde donors, serving as preservatives that slowly release formaldehyde as they break down over time. Check out the building material's safety data sheet or ingredient list to see what ingredients are included in the product you're researching. These formaldehyde donors have a long list of names, and many are found in cosmetics as well, such as DMDM hydantoin, diazolidinyl urea, and sodium hydroxyl, just to name a few.

Then there are products made from recycled newspaper or plastic. Recycled detergent bottles can give your home a perpetual smell of synthetic fragrances, while recycled paper products can off-gas VOCs from the xylene found in inks. Although we fully support sustainability and eco-conscious efforts, our number one priority is the well-being of the people and animals that reside within a home.

Homeowners often come to us already frustrated after reading books by or working with consultants who claim to be experts in healthy building but led them down an expensive path full of mistakes and misinformation. As the founders of JS2 Partners and Healthier Homes, we have a combined thirty years of building and design experience with construction projects ranging from multimillion-dollar commercial buildings to fully custom residential homes. It's one thing to talk about the products you'd recommend for creating a healthy environment but quite another to actually implement novel healthy-home building practices.

Another common question is, how much more will healthy home building cost? Most people are surprised to learn that our materials and methods are typically no more expensive than those offered by other high-quality custom home builders. The difference lies within our level of materials knowledge, refined methods, and years of research and implementation. Even so, there is truth to the phrase "you get what you pay for." High-quality materials and craftsmanship should be a priority when making one of the largest investments in your lifetime.

BETTER MATERIALS, PROVEN METHODS

We've done extensive research on the most popular conventional building methods and delved deeply into alternative building practices, too, including straw bales, hemp-based plaster, magnesium oxide products, and clay. While some of these products are clean and made from high-quality ingredients, others contain waste materials or are manufactured in overseas factories with no way to trace their ingredients. A fair number of alternative-type materials have resulted in major structural defects and poor performance when put to the test, and others don't meet our stringent requirements for healthy materials.

For these reasons and more, we only provide building material recommendations and methodologies that have proven safe and effective in the field. We look not only at the constituents of these building materials but also at how the build-out is deployed. Our healthy homes are proof that it is possible to construct a beautiful, functional, and nontoxic house using modern materials that won't break the bank. In fact, our homes would fit into any neighborhood, and no one would ever know they were constructed differently unless you told them. But, as always, our number one priority is occupant health and well-being.

Have we made a few mistakes along the way? Sure. But with Jen's decade of extensive product knowledge and keen eye for design and Rusty's twenty years of firsthand commercial and residential building experience, we have the know-how to get creative when we need to. Rest assured, if we share it in this book, it's a proven method we've used in homes we have built. This book is designed to make your life easier, not harder.

That said, healthy building products, methods, and technologies are constantly evolving, which is why we created a place where you can access the most accurate and up-to-date information. The Healthier Homes online community was designed to serve as a dynamic resource for all things healthy home living and includes a shopping section with nontoxic construction materials and curated home furnishings. We also offer eGuides that expand on several important topics mentioned in this book, letting you choose which areas you'd like to explore further.

But this book is where it all begins, and it will form your foundation for a healthier life. In the following pages, we share our building experience and explain how to use safe, practical materials and methods to construct the healthy home of your dreams. Our goal is to provide a safe, comfortable, and beautiful place for homeowners to lovingly call home.

HOW WE APPROACH TRIAL AND ERROR

In construction, methods and materials are often proven effective through their use in the field and by standing the test of time. We've learned a great deal about healthy building through years of trial and error, even in our own home. We're an interesting case compared to other builders—part science geeks, part human health experts, and all builders. If we haven't tried it or seen it used successfully in other projects, then we conduct controlled testing on materials and implementation practices before trying them in the homes we build.

There's a reason professionals like general contractors and builders exist. Each construction material, installation method, and building environment is unique. We admire DIYers for their initiative, but we also ask them to respect the complexity of building. Some healthy home recommendations you might find online could be misguided or dangerous. But, if a method is out there, chances are our team has tried it. If not, we will find a safe way to verify whether a new idea will work. There's often more than one solution to home-building challenges, which is important knowledge because each project has its own set of unique factors.

Before you decide to embark on your own healthy home experiment, check the Healthier Homes website, where we are constantly adding how-to guides, articles, and videos covering all things healthy home building.

HealthierHomes.com

key points

▸ Our primary goal is promoting occupant health and wellness.

▸ This book is designed to be easy to read and understand, no matter your experience level.

▸ Stress is a part of life, but your home should be a place of rest and rejuvenation.

▸ Modern construction methods and products may stress the nervous system and inhibit rest and healing. These include everything from formaldehyde in lumber, flooring, walls, paints, and furnishings to wireless signals from smart home technologies.

▸ Combating those negative effects starts with education.

▸ Healthy homes are not necessarily "green"; instead, they are built with structurally sound materials and methods that promote occupant wellness.

▸ With our help and this book, you can have a beautiful, functional, and nontoxic house without breaking the bank.

▸ Visit HealthierHomes.com for more information and product recommendations.

Next up → start thinking about your own indoor living environment with a quick and handy checklist of healthy living priorities.

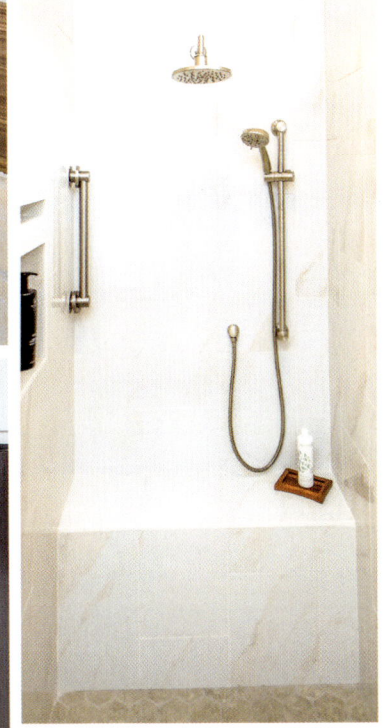

evaluating your indoor living environment

We are a design-build construction company, meaning our team works with homeowners from the beginning of the design process through the completion of construction.

Careful planning at the beginning of a home project, before addressing anything in the field, will help prevent costly mistakes or changes down the road. So, let's use the following lists of questions to evaluate your wants and needs for your new space. Then, at the end, you can assess the functionality of your current home. After all, knowing what you hope to accomplish with a new home (or a renovation) will put the information in the rest of this book into perspective. You can also use these lists as sources of questions for your builder.

FUTURE LIVING SPACE ASSESSMENT

KITCHEN:

Do you cook often? Yes ☐ / No ☐

Do you prefer gas or electric appliances? ....

Do you want a double oven? Yes ☐ / No ☐

How often do you use a microwave? ....

What kind of cooking ventilation system do you prefer? Where should it vent to?
How often do you wash the filter media?
....
....
....

What are your storage needs for refrigerated food, frozen food, beverages, and
canned and dry goods?
....
....
....

What are your storage needs for pots and pans, bakeware, china, everyday
dishes, silverware, and glassware?
....
....
....

Do you want to keep countertop clutter to a minimum with a pantry large
enough for food, appliances, vitamins, pet supplies, etc.? Yes ☐ / No ☐

How many drawers and cabinets do you need? ....
....
....

Do you want a separate prep space with outlets for a coffee maker, blender,
and/or electric kettle? Yes ☐ / No ☐

Do you need a space to put a water filtration system or water dispenser? Yes ☐ / No ☐

Do you want your refrigerator, oven, and sink in close proximity? Yes ☐ / No ☐

Do you want seating at an island or countertop so that people can congregate in the kitchen? Yes ☐ / No ☐

How important to you are natural light sources? ....
....
....
....

What type of countertops do you prefer? (Consider what is functional, attractive, and easy to clean without harsh chemicals or sealers.) ....
....
....
....

GARAGE:

How many cars do you own or plan to own? ....

How many kids (and toys) do you have? ....

Do you need a workshop?   Yes ☐ / No ☐

Do you want extra storage for outdoor adventure equipment, lawn
equipment, sports equipment, tools, etc.?   Yes ☐ / No ☐

Do you want a second refrigerator in the garage?   Yes ☐ / No ☐

Do you want a mudroom or bench area in the entryway where you can hang
jackets, drop work bags, and store shoes?   Yes ☐ / No ☐

OFFICE:

Do you work from home?   Yes ☐ / No ☐

Do you have clients or contractors coming in and out of your home?   Yes ☐ / No ☐

Do you have a lot of computer equipment or books?   Yes ☐ / No ☐

Do you have a need for floor outlets?   Yes ☐ / No ☐

Should the office serve as a flex space for hobbies such as arts and crafts?   Yes ☐ / No ☐

Do you need adequate meeting space with A/V capabilities?   Yes ☐ / No ☐

LIVING/DINING:

How big are your dining room table and chairs? ....

Do you entertain often? Yes ☐ / No ☐

Do you prefer an open concept, meaning a layout that emphasizes one large open space? Yes ☐ / No ☐

Do you like having floor outlets for lamps or plugging in a laptop on the couch? Yes ☐ / No ☐

Do you want a fireplace? Yes ☐ / No ☐

If yes, does the fireplace need to provide heat? Yes ☐ / No ☐

How extensive are your multimedia needs (TV, stereo system, etc.)? ....

Do you need storage and organization for kids' toys? Yes ☐ / No ☐

LAUNDRY:

How big is your family? More people equal more laundry. ....

How often do you do laundry? ....

Do you want the laundry room to be upstairs or downstairs? ....

How often do you hang clothes to dry? ....

Do you put pet food bowls or litter boxes in the laundry area? Yes ☐ / No ☐

Do you want a second refrigerator or freezer in the laundry area? Yes ☐ / No ☐

How close to the bedrooms do you want your laundry room to be? ....

MASTER SUITE:

How big is your bed? ....

How much room do you need for nightstands and dressers? ....

Do you have lots of clothes, shoes, and bags?   Yes ☐ / No ☐

Do you want double closets, or will a shared closet suffice? ....

Do you spend a lot of time in your bedroom?   Yes ☐ / No ☐

Do you want a separate shower and stand-alone tub?   Yes ☐ / No ☐

How much vanity space do you want? ....

Do you need plenty of linen storage?   Yes ☐ / No ☐

How much natural light suits you? ....

Do you want your toilet in a "water closet"?   Yes ☐ / No ☐

Is security a concern?   Yes ☐ / No ☐

OTHER BEDROOMS AND BATHROOMS:

Do you live with family or plan for guests to stay often?   Yes ☐ / No ☐

Are you OK with occupants sharing a bathroom, or do they each need their own? ....

Do these bathrooms need showers or bathtub/shower combos? ....

Should these bathrooms also have linen storage?   Yes ☐ / No ☐

Will you need a nursery in the future?   Yes ☐ / No ☐

How close to the master suite do you want the nursery to be? ....

How close to the master suite do you want the other bedrooms to be? ....

BONUS SPACES:

Do you work out often and/or have lots of exercise equipment?   Yes ☐ / No ☐

Do you have a morning yoga routine?   Yes ☐ / No ☐

Do your kids enjoy outdoor sports (e.g., do you need a basketball area on the driveway or extra play space in the backyard)?   Yes ☐ / No ☐

Do you want to have occasional get-togethers around a table, such as game night or book club?   Yes ☐ / No ☐

Do you want a pool and/or hot tub?   Yes ☐ / No ☐

Do you want an infrared detox sauna?   Yes ☐ / No ☐

Do your kids need a playroom?   Yes ☐ / No ☐

Do you like to garden?   Yes ☐ / No ☐

Where do you want your patio and deck areas? ....

Do you grill or want an outdoor kitchen?   Yes ☐ / No ☐

What makes you feel safe at home? ....

Do you want extra storage for holiday decorations, coats, luggage, etc.?   Yes ☐ / No ☐

CURRENT LIVING SPACE ASSESSMENT*

Do you feel tired when you wake up in the morning?	Yes ☐ / No ☐
Are there areas in your home that make you feel sleepy or jittery?	Yes ☐ / No ☐
Do you find it hard to focus and/or be productive at home?	Yes ☐ / No ☐
Do you consider your house cluttered?	Yes ☐ / No ☐
Is it hard for you to keep your house clean?	Yes ☐ / No ☐
Are there places in your home that always seem damp or smell musty?	Yes ☐ / No ☐
Do areas in your home feel drafty?	Yes ☐ / No ☐
Are there cracks in the walls or ceilings?	Yes ☐ / No ☐
Is the humidity in your home constantly above 50 percent?	Yes ☐ / No ☐
Is the paint or wallpaper peeling?	Yes ☐ / No ☐
Does your tap water have a color or an odor?	Yes ☐ / No ☐
Has your house ever had a leak?	Yes ☐ / No ☐
Does water pool around the foundation in certain areas outside?	Yes ☐ / No ☐

Are your floors uneven or warped in random spots? — Yes ☐ / No ☐

Are your floorboards or window trims buckling, bubbling, or peeling? — Yes ☐ / No ☐

Do your lights or electrical wires buzz or flicker? — Yes ☐ / No ☐

Do you leave your wireless router on 24/7? — Yes ☐ / No ☐

Do you use smart home technology? — Yes ☐ / No ☐

Do you notice areas of condensation inside the house? — Yes ☐ / No ☐

Does your home have that "new car smell"? — Yes ☐ / No ☐

Does your home smell like pepper or plastic? — Yes ☐ / No ☐

Does your house smell like an old book? — Yes ☐ / No ☐

Do you ever smell sewage odors inside? — Yes ☐ / No ☐

Does your house smell like paint fumes or car exhaust? — Yes ☐ / No ☐

Have you noticed stains on your walls or ceilings? — Yes ☐ / No ☐

Do you ever hear water dripping behind a wall or in the fireplace? — Yes ☐ / No ☐

If you answered yes to any of these questions, you may have cause for concern regarding your home's toxicity levels. Not to worry—we will address how to correct these issues and more in upcoming chapters.

Let's move on to our next topic. What better way to segue into how to build a healthy home? First things first: choosing a healthy place to live.

choosing a healthy place to live

LOCATION, LOCATION, LOCATION

The first step toward creating a healthy living space is to choose a geographic location with the least amount of pollution possible. Many of us overlook factors such as noise and air pollution, EMFs, and pesticide use, all of which have become increasingly commonplace in our modern world.

Although it would be nice to live in the country and remove pollution from our living situations entirely, our lifestyles often demand we stay close to workplaces, schools, grocery stores, and other necessities. Plus, a sense of community and an urban environment can be very appealing. Whole Foods is at the top of our must-have proximity list! What's on yours?

Air pollution should be at the top of everyone's avoidance list. Obviously, it's important to choose a building site that is not downwind of refineries, factories, feed lots, conventional farms, and major interstates, but more goes into it than that.

For example, Jen lived in Dallas during her twenties, and although her apartment was in the lively and hip uptown area, it was also in the direct path of the landing strips at Love Field, an international airport. She later discovered elevated levels of xylene and ethylbenzene in her blood—both of which are found in jet fuel exhaust.

According to the EPA, ethylbenzene and xylene are hazardous VOCs. Long-term exposure can cause respiratory issues and nervous system disorders and has been linked to an increased risk of kidney, liver, lung, and testicular cancers. Sadly, you can't escape these offenders simply by avoiding airports. Synthetic fragrances, industrial emissions, and building products such as paints, solvents, and coatings also contain high levels of these chemicals.

DREAMING OF COUNTRY LIVING

At this point, you may be considering a move to a rural area to eliminate air pollution from industrial and transportation sources. While that could help, country living comes with other airborne contaminants.

Over the past decade, consumers have demanded more organic and non-GMO foods. The Healthier Homes team has embraced the benefits of an organic diet, too; it makes us feel more energized and reduces our toxic load, and we're all about promoting responsible farming and ranching practices.

But many crop farmers receive government subsidies that promote conventional farming practices such as the use of GMO seeds and toxic pesticides. One example of these products is the popular weed killer Roundup, for which manufacturer Monsanto has come under scrutiny after class action lawsuits linked Roundup to thousands of cases of non-Hodgkin lymphoma.

This may leave you asking what's in your steamed veggies—or, better yet, thinking twice before moving to an area that used to be conventional farmland.

Because bugs, fungal infections, and bacterial overgrowth can wipe out entire fields of crops in a matter of days, farmers who are not trained in organic practices often have no choice but to use toxic biocides on a massive scale to protect their livelihoods. So, if you're considering purchasing land near a farm, do your homework and find out what kind of farm your new neighbors run.

NEVER USE ROUNDUP OR SYNTHETIC HERBICIDES

Roundup is one of the world's most popular weed killers, but it contains glyphosate and other chemicals that can damage human cells, potentially causing cancer and other health problems. If you love gardening and want a healthy alternative, follow our recipe for homemade weed killer below.

Homemade weed killer ingredients:

- ½ gallon strong vinegar (at least 20 percent acetic acid content)

- 1 tablespoon dish soap (we recommend using an all-natural plant-based brand)

- ½ cup salt (optional)

Mix the vinegar and dish soap in a spray bottle by shaking and spray the mixture on weeds, carefully avoiding other plants. The acetic acid in the vinegar will kill weeds on contact, while the soap acts as a surfactant, helping the vinegar stick to the desired area. (Only add the salt if you never want plants to grow where you spray this mixture.)

HOW TO PERFORM A SOIL TEST

- **Investigate the property's history:** Talk to the previous owners and visit the city archives to learn the history of your property and the neighboring area. If the soil has a history of contamination, find out what remediation techniques were used.

- **Find a lab:** Ask around (and go online) to find a reputable soil test lab near you. Make sure they're accredited and find out what they test for. Also ask for special instructions for handling and sending soil samples. Prices and turnaround times may vary, so inquire about those as well.

- **Gather samples:** Buy a soil kit or gather your own supplies as instructed by the lab, which may include a shovel or trowel, plastic buckets, resealable bags, and paper and pencil. You'll probably need to walk your lot, collecting samples from various locations (and recording them). In each spot, you may need to remove vegetation and dig sixteen inches deep before extracting a soil sample.

Performing a soil test for contaminants is also a good idea (one may be purchased through some university agriculture programs, or you can hire a soil engineer). If the farm is not following organic practices, the surrounding area's groundwater and air will likely be contaminated with pesticides, herbicides, and organophosphates. The same concern goes for purchasing land in new developments, many of which were once conventional farms, meaning the land is likely contaminated. And while living near a country club can be prestigious, golf courses tend to use large amounts of pesticides as well.

Another equally important kind of soil test ensures the soil on your lot can support a home. Soil plays a huge role in the building process and the stability of the home years after it's built. If the soil is loose or easily compressed, your foundation may need extra supports. As such, it's often a good idea to work with a soil engineer.

Elevation can also affect your living space. Houses built higher up and outside of valleys tend to have cleaner air since pollution and smog typically settle in lower areas. One of our clients in Las Vegas, for example, looked up the pollution height in their area and then purchased a hillside lot just above the local 700-foot smog delineation level.

Many think that mountainous regions such as Colorado have some of the cleanest air possible. This may be true if you're living in the Rockies, but Denver's brown cloud is nasty and real. This is partly attributed to the fact that gasoline engines run less efficiently at higher altitudes and emit greater amounts of hydrocarbon and carbon monoxide pollution. Reduced oxygen levels coupled with higher levels of photochemical smog can have a significant negative effect on human health.[1]

For more information about seasonal air pollution in the areas you're considering, look up the EPA's Air Quality Index Report, which provides annual air pollution information for major cities in the US, including the number of days when the air was very unhealthy or hazardous. Just keep in mind that these statistics are not finalized until May of the following year.

WE ARE ELECTRICAL BEINGS

We've all heard the joke about living under power lines, but chronic exposure to EMFs, especially those from transmission lines, is another dangerous factor to consider. Humans are basically walking bags of water (it makes up 45 to 75 percent of our bodies), and water conducts electricity. In fact, that's how our nervous systems work—by sending electrical signals from our brains to different areas of our bodies.

Interference from outside sources can cause the electrical parts of our bodies to start to go haywire. Everyone is affected by electrical interference to some degree, but certain people develop heightened sensitivities to them. These individuals often become aware of the presence of these frequencies after suffering adverse physical reactions. Excessive EMFs can cause heart rate increases as well as systemic inflammation. However, the body's defensive response to EMFs happens for a reason. More on that shortly.

GEEK BOX | **what are EMFs?**

In the simplest terms possible, an electromagnetic field (EMF) is a field—or a region of space affected by a force—created by an accelerating electric charge. An EMF is the invisible energy radiating from cell phones, microwaves, and other devices.

The scientific community is still uncovering the effects EMFs have on human health, but we are electrical beings, and evidence suggests EMF exposure can lead to the following:

    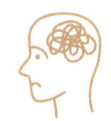 

SLEEP DISTURBANCES   HEADACHES   DEPRESSION   FATIGUE   DIFFICULTY CONCENTRATING   MORE SERIOUS HEALTH CONDITIONS, INCLUDING CANCER

An EMF meter is a must when looking at a home or property to purchase. Some power lines produce much larger and/or stronger fields than others, and using a meter is the only way to know if your future home resides inside one of these EMF fields.

When looking for an EMF meter, consider the following question: what type of field do you want to measure? EMFs come in different varieties, including magnetic, electric, radio, and harmonic. Electric fields, for example, come from power lines and wiring, while radio and harmonic fields are the flavor of Wi-Fi and Bluetooth. Some meters measure multiple types; these tend to be more expensive but might suit your needs better. Also keep in mind that these meters are complex devices, so be sure to follow the manufacturer's instructions to ensure accurate readings.

Some metropolitan areas have such concentrated EMFs that the ground itself becomes saturated with electricity. The ground normally serves as a medium for, well, grounding out these fields, but eventually there's not enough ground to keep up with all the electricity. This phenomenon is known as EMF saturation,[2] and it can happen in the ground or in the space around us. This is another good reason to seek areas to live with fewer people, businesses, and office buildings.

Stay away from cell phone and satellite transmission towers as well. We cringe when we see that the owner of an office building has set up cell phone antennas to earn an extra buck at the expense of the workers inside, who spend eight-plus hours a day being exposed to constant microwave radiation.

WIRELESS RADIATION: EFFECTS AND SOURCES

Radio frequencies have increasingly permeated our living spaces in the past fifty years or so, and the new generation of more powerful Bluetooth, 5G, and Wi-Fi bands has drastically proliferated inside our homes over the past decade. These powerful frequencies fall at the microwave end of the EMF spectrum.

As these technologies began to emerge, they enabled amazing new communication methods that put entire libraries of information at our fingertips. However, no studies were conducted on the safety of living near concentrated sources of wireless radiation. Over time, a growing body of evidence has begun to suggest that chronic wireless radiation exposure from living and working in EMF-saturated environments may be causing non-thermal ill health effects such as hair loss, asthma, allergies, headaches, chronic pain, difficulty concentrating, and skin and eye problems.[2]

So, why in the world are wireless technologies considered OK for daily human exposure? Truth is, they aren't.

A 2006 Swedish survey suggested that between 230,000 and 290,000 people showcased various symptoms when exposed to EMF sources. As a result, Sweden officially recognized electrohypersensitivity as a functional impairment (not a disease). Furthermore, epidemiological studies pointed to a correlation between cancer and long-term exposure to magnetic fields and microwaves.[3]

Despite this evidence, not only cell phones but also computers, internet routers, A/V equipment, appliances, and even baby monitors now operate wirelessly. Congested neighborhoods, high-rises, and office buildings have become mega-sources of radiation. We recommend finding a place to call home that gives your body a break from the electrosmog.

Don't worry—we're not suggesting everyone should go back to the Stone Age. We love our cell phones and computers, too. But there are ways to be safe about EMF sources, such as turning off wireless technology within the home at night. We'll discuss these precautions in greater detail later in the book.

OFTEN OVERLOOKED DISRUPTIONS TO HEALTHY LIVING

Noise is another form of pollution for humans and animals. Although it might "sound" ridiculous, constant or frequent noise and vibrations can put stress on the body. And, as we discussed earlier, constant stress prevents the body from entering the rest-and-repair phase. Over time, a stressed body that isn't allowed to heal will break down and fall ill. Living close to an airport or a busy road can be a source of constant background noise as well as fumes.

Also check to see if your city and/or property owners association has zoning requirements. Otherwise, you may wish you had—especially when your next-door neighbor opens a noisy motorcycle repair shop right after you finish building your new home.

Climate is an important consideration as well. Managing indoor moisture in tropical and rainy regions tends to be challenging, but proper healthy building practices make it possible to keep mold and mildew problems at bay.

Take a look around, too. Consider your lot and its relevance to the surrounding land. Living downhill from a soon-to-be developed area may lead to a major shift in the watershed. Or, if you love a property with a view, keep in mind that future building may end up blocking the scenery.

You should also be aware of regional energy codes. Builders are required by cities, counties, and local code requirements to adhere to minimum energy standards, which can be a substantial cost driver for construction materials and methods that must meet codes in a given region.

Bottom line: we spend a lot of time at work, in school, and on the road, where our bodies are exposed to pollutants that we cannot control. Our homes should promote rest and relaxation to mitigate the fight-or-flight stress response. Choosing a healthy place to live is all about reducing stress on your body by making smart decisions when it comes to geographic and environmental factors.

Before choosing your location, however, you should meet with your builder or general contractor. That way, you can involve them in the design process from the start.

why do people live longer in certain parts of the world?

Various factors may influence the regional impact on life expectancy. Parts of the world where people tend to live longer are often called "Blue Zones," and they tend to have several traits in common, including cultural norms relating to a healthy diet and lots of physical activity.

One interesting and often overlooked factor of Blue Zones, however, is that they typically occur in less densely populated and somewhat isolated communities. Blue Zones that are far from major sources of pollution include the island of Ikaria in Greece and the Nicoya Peninsula of Costa Rica. We wonder if other factors such as less noise, fewer sources of EMFs, and a reliance on open-air ventilated buildings might play a role as well.

key points

▸ Location is an important factor when it comes to building a healthy home.

▸ Living near non-organic farms, busy airports and highways, or heavily populated urban areas is often a bad idea.

▸ Air pollution exposes your body to a variety of toxic chemicals, but choosing a home at a higher elevation may help.

▸ Avoiding electrosmog, or radiation from EMFs, may reduce your risk of cancer and other health problems.

▸ Noise pollution can keep your body in a high-stress state so that it rests and heals less efficiently.

▸ A variety of geographic and environmental factors can affect your living environment and health, so choose carefully—but not before contacting your builder.

Up next → get the lowdown on how to design your living space with health in mind.

HEALTHY HOME by JS2P

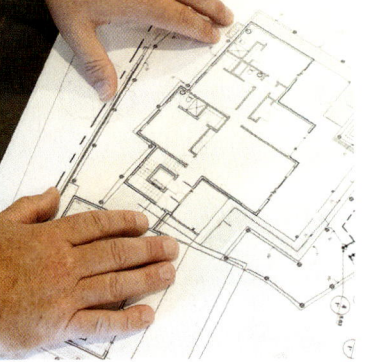

designing a healthy living space

We always recommend bringing your builder or general contractor (GC) in on a project before making any big decisions—from choosing your location to designing your home.

Insights from your builder can help determine cost drivers that will affect your design decisions and building expenses down the line. Factors such as rocky soil, slopes, drainage from neighboring properties, local energy codes, and community requirements are just a few to consider.

Think of your builder or GC as a valuable team member for your construction project—you should feel confident in their knowledge and experience. If they balk at or seem annoyed by the idea of using new products or slightly modified building methods to eliminate toxicity, this should be a red flag. A small amount of initial hesitation is natural, but your builder should ultimately be on board as an advocate for your new healthy home. If you don't get that vibe, find a builder you trust.

While looking for a builder, you should first be aware of the different types of people who may be involved in the process. Here are a few of the most noteworthy:

ARCHITECT:
This is the person who turns your big ideas into a workable blueprint. They may also be called a drafter. Many people hire an architect to design a home and then a general contractor or builder to implement the plan.

GENERAL CONTRACTOR (GC):
The GC oversees various types of construction projects inside of a home and is usually hired for remodels.

CUSTOM BUILDER:
A custom home builder (like us) oversees every part of the new home build process, from site preparation to final landscaping, meaning you don't need to worry about finding anyone else.

When you start a conversation with a custom builder, it's important to have a few things in mind to ensure you're on the same page. First, we recommend making a list of all your must-have and wish-list features and amenities. Obviously, if you're reading this book, healthy construction is one of them, but be sure to go into detail. (This book should make it easier to build that list.) Also remember to check credentials and verify experience before signing a contract. Finally, ask plenty of questions related to the things you care about. As previously mentioned, if a builder seems dismissive of your healthy home concerns, it might be best to keep looking.

So, you've found the perfect builder or GC and the perfect location to build on. Each piece fits your budget, checks the important boxes, and will provide a healthy and safe home environment for you and your family for years to come.

Now what?

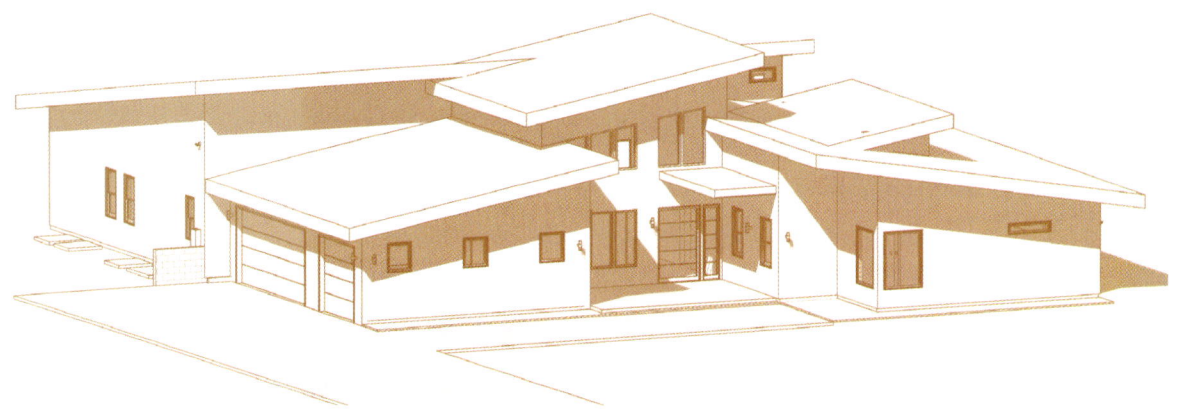

FORM, FUNCTION, AND...COMPOSITION?

Builders, designers, and architects all acknowledge that form and function are concepts that affect the home drawing and building process. But what do those words mean in this context?

There are two ways to look at the relationship. For some, form is the "how" and function is the "why." We consider *how* to build the house and *why* to build it that way. As such, form should always follow function. There's also a school of thought that approaches form as the beauty of a structure and function as, well, the functionality of the space.

Take the example of a garage floor. Consider a garage floor finished with carpet. From a functionality standpoint, laying carpet in an unconditioned space that gets hot, wet, and dirty from cars being parked on it doesn't make sense. Most garage floors need to withstand moisture, dirt, a wide range of temperatures, and weight from heavy vehicles.

This example might seem silly or obvious, but it illustrates the point that form (the choice of flooring) must follow function (designing the space to handle vehicles and outdoor elements). But our building philosophy doesn't end there.

As healthy home builders, we add a novel third component to form and function. This third element is *composition*.

Composition is the "what"—what these materials are made of and what impact they have on occupant health. A home can be beautiful and functional while being healthy and safe, which is the point of composition. We thoroughly vet the composition of all construction materials that go into our homes, and you should, too.

CREATING A FUNCTIONAL FLOOR PLAN

Now it's time to consider home layout plans that fit your square footage needs, budget, and property constraints. Remember, your layout should suit your lifestyle now while supporting potential changes in the future.

Floor plans with open layouts tend to circulate air better and appeal to most people. An open layout is especially important if you enjoy entertaining guests or are considering resale value.

Another factor is whether you are planning for little ones. Maybe you need a room near the master bedroom that you can convert into a nursery for a new baby. Or perhaps you already have children and want to keep the master suite away from the kids' bedrooms to help reduce noise in boisterous households.

Placing the laundry room on the same floor as most of the bedrooms helps decrease the number of trips taken up and down the stairs with heavy baskets in multistory homes. We've designed homes for several clients with two sets of washers and dryers—one set in the actual laundry room and the other in the master closet.

Every household has unique storage needs. If you're an aspiring chef who owns lots of kitchen gadgets or a vitamin and supplements aficionado (like us), adding extra storage in the kitchen and pantry will be beneficial. A wine collector or someone who entertains often may want to consider adding a small bar or incorporating storage space for dishes and beverages. Maybe you are the artsy type and need a small closet for your craft supplies and paints. No matter your jam, incorporating strategically placed storage is always a good idea.

Also think about whether this is going to be your forever home, which may have to accommodate your changing needs as your family matures. We builders often refer to this concept as "aging in place." We design the doorways wide enough to accommodate a wheelchair and often strategically place a storage closet so that it can one day serve as an elevator shaft. A little foresight can go a long way.

Another often overlooked factor is the space between the kitchen island and the surrounding cabinets. Be sure your refrigerator, oven doors, dishwasher, and drawers all have adequate clearance to open and allow traffic to pass by. Also, planning for some additional storage space in the garage for toys and tools (and even items like bikes and a lawnmower) is important.

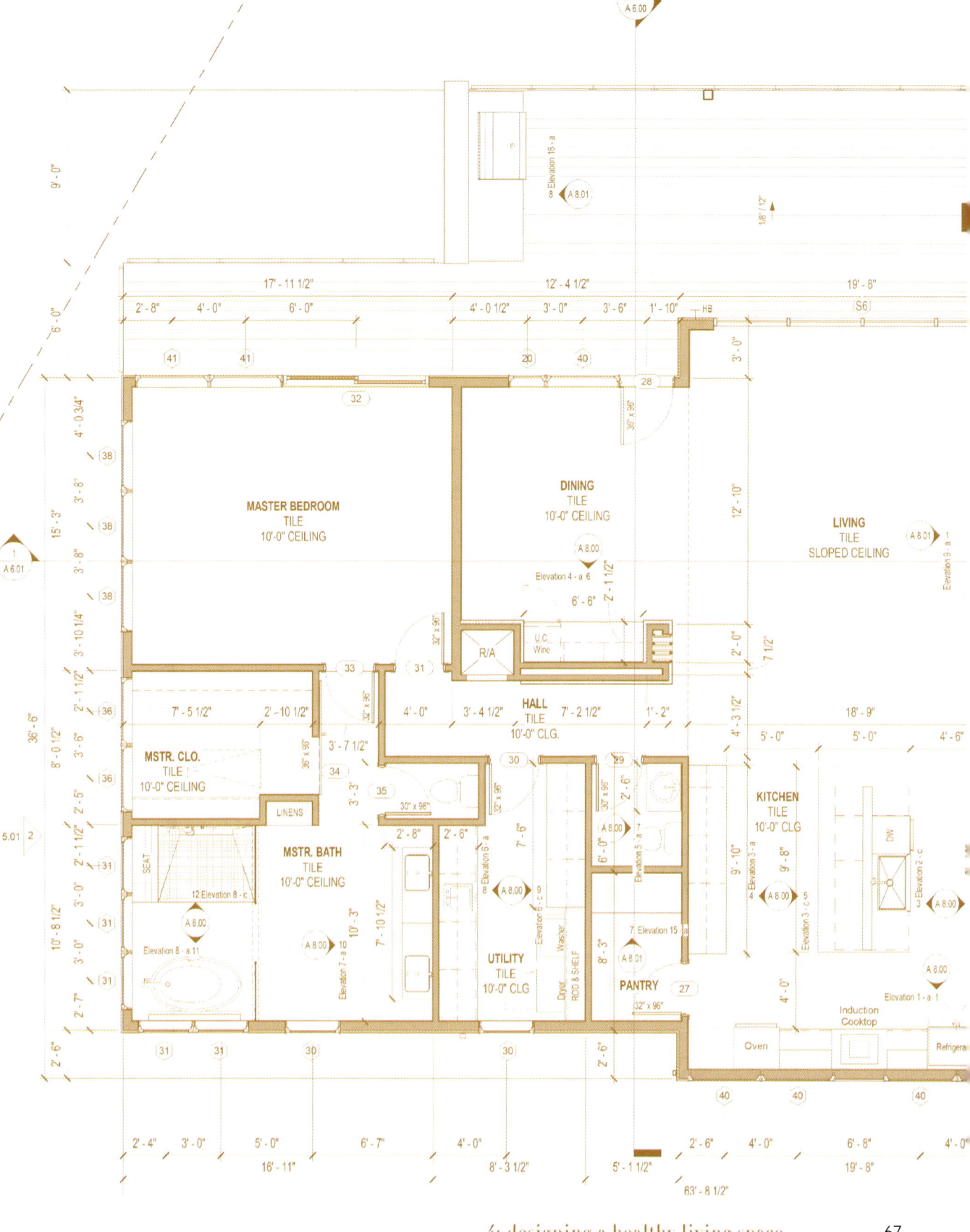

MASTER BEDROOM
TILE
10'-0" CEILING

DINING
TILE
10'-0" CEILING

LIVING
TILE
SLOPED CEILING

MSTR. CLO.
TILE
10'-0" CEILING

LINENS

MSTR. BATH
TILE
10'-0" CEILING

HALL
TILE
10'-0" CLG.

KITCHEN
TILE
10'-0" CLG

UTILITY
TILE
10'-0" CLG.

PANTRY

SEAT

U.C.
Wine

R/A

Dryer Washer

ROD & SHELF

Oven

Induction
Cooktop

Refrigera

PLOTTING YOUR PROPERTY

Creating a plotted plan means designing the layout of the home on your property. This should happen early in the design process, preferably as soon as you know the desired square footage, settle on a floor plan, and have a professional survey done.

Neighbor proximity is also important to consider. If one of your neighbors has a noisy workshop or large satellite dish, you may want to place your home farther away from that house. If you live in a rural area without trash pickup, avoid placing your home close to your neighbors' burn piles. Also consider where laundry exhaust vents are located. We personally don't enjoy smelling our neighbors' laundry detergent as we relax on the patio with an early-evening drink.

If your lot backs up to a busy street, position your home away from traffic noise, and avoid putting fresh air intakes on the sides of the home that will have direct exposure to exhaust fumes from passing vehicles.

ESSENTIAL AMENITIES, INSIDE AND OUT

A sauna is an excellent way to detox and keep your cardiovascular and immune systems in tip-top shape. If you want one, we recommend planning a space in the master bathroom to accommodate a sauna. Don't forget to place an extra circuit on the electrical plans to cover the amps required for your unit(s).

It's also essential to have a place to relax in the fresh air and get your vitamin D from the sun. Plan for patios or decks that receive ample sunlight throughout the year. Although the amount of sunlight depends on your geographic location and the season (winter being the worst), solar noon is the best time to get some sun no matter where you live.

Placing your home according to the orientation of the sun at different times of the year can be advantageous for both energy savings and thermal comfort. Situating outdoor living spaces in areas conducive to ample air flow helps prevent stagnant heat and standing water issues. No one wants a mosquito pond next to their patio.

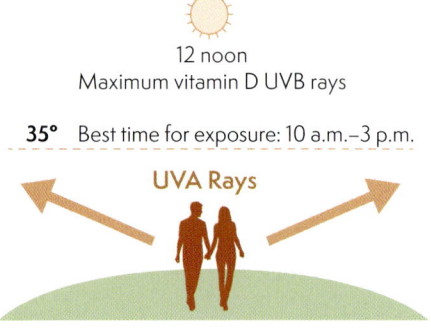

12 noon
Maximum vitamin D UVB rays

35°   Best time for exposure: 10 a.m.–3 p.m.

UVA Rays

WHAT YOUR PROFESSIONAL SURVEY SHOULD INCLUDE

Hiring a licensed survey company to plot your land is an important part of the design framework process. A licensed surveyor will locate the underground iron pins that verify your property's metes and bounds (the physical features and boundaries of a piece of property). This is important not only for verifying your purchase but also for ensuring your drafting team has the info they need to design and plot your home. Be sure to ask your surveyor to include

- **Cardinal points:** North, south, east, and west.

- **Topographical details:** These tell your builder and drafter where the slopes are on your property along with the precise measurements of the land's elevation changes.*

- **Tree survey:** This records the locations, trunk caliper sizes, and species of trees on your property. Keep in mind that some neighborhood association guidelines require homeowners to replace trees that are cut down with trees planted elsewhere on the property.

- **Setbacks and utility rights-of-way:** These lines dictate where your builder is allowed to place the home's building envelope on the property.

If your property has a fair amount of slope, it may be worth bringing a civil engineer on board as well. Building on the side of a hill presents challenges to ensure proper drainage. This is money well spent when it comes to healthy building. We will cover drainage in more detail later in the book.

WORK AND PLAY

Do you work from home? A home office should provide natural light and a place for concentration away from distractions. Floor plugs require wiring that goes into the foundation and should be planned for in advance. You should also place laundry rooms, gyms, and home theaters away from offices, bedrooms, or any other areas where noise and vibrations may interfere with those trying to concentrate or get some shut-eye.

What are your outdoor hobbies? If you love sports, consider making room to install a putting green or volleyball net in the backyard. If you love to cook, maybe include an outdoor kitchen in your budget—just be careful to place your grill away from your HVAC system's fresh air intakes.

Swimming is excellent cardiovascular exercise, and a pool is a fabulous place to hang out with friends and family. We used to recommend saltwater pools over the heavily chlorinated kind, but UV light and ozone technologies allow pools to remain in balance while drastically reducing the amount of chemicals used to keep bacteria and fungi in check. (Saltwater pools are still wonderful for the skin, but the equipment tends to corrode much quicker.) More on these later.

PARKING AND SECURITY

Ample parking and driveway space are usually at the top of all homeowners' wish lists. There are ways to get creative with your lot's topography to allow more space for vehicles. For example, if you want more yard without sacrificing parking, one clever idea is to place cinder blocks in a parking area and fill them with soil and grass overseed. This makes for a stable and sturdy parking area disguised by grass.

Feelings of safety and security are important for all homeowners. If you live in a large city, your neighborhood may experience a higher risk for crime. Perhaps a gated community would make you feel more secure. Whatever you decide, you should budget and plan for a fence, your own entry gate, and/or an alarm system during the design phase of a build.

Other healthy home safety and security considerations may include investing in steel doors, using longer screws to secure the strike plate and hinges, and buying high-security locks. You can also install security cameras or motion-operated lights to complement your security system, but make sure none of these components are wireless (or that wireless capabilities can be disabled) if you're looking to avoid EMFs.

THE PASSIVE HOUSE MODEL

The building science community's passive house model advocates building homes for maximum energy efficiency with minimal ecological impact. We deploy several of these concepts in our design-build model, and some are achievable in renovation projects, too.

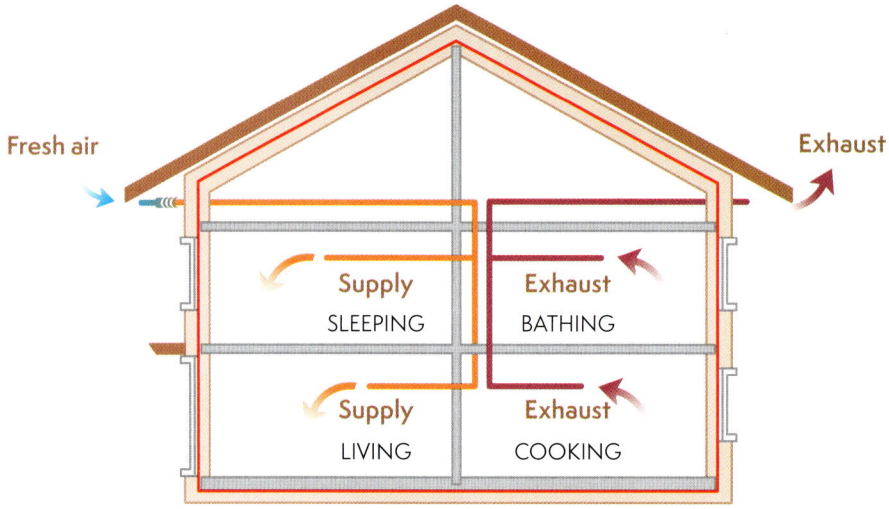

Solar gain is an important factor. It refers to the way a structure's thermal energy increases as it absorbs radiation from the sun. Controlling solar gain within the home helps lower cooling bills during the summer months. If your lot allows for it, you should situate your home so that the windows in living spaces used during the day get direct sunlight during the winter months.

Also pay attention to wind patterns, which usually prevail from different directions according to the season. For example, if you live on top of a hill that faces north, winter cold fronts are likely to bring strong winds and driving rain on that side of the property. In that scenario, it is essential to use proper water intrusion mitigation techniques and ensure the windows, doors, and roofing allow water to drain away from the home. Routing water away from the exterior is key, especially considering the path may change depending on which direction the rain is coming from.

HOW WE GO THE EXTRA MILE

To ensure energy conservation and occupant health, we build homes that exceed standard energy code requirements. For example, our building envelopes—the physical separators between the conditioned and unconditioned spaces of a house—are always airtight, regardless of whether local building codes require it. We also source commercial-grade insulation that performs significantly better in both hot and cold climates, and we use doors and windows that have a true thermal break with insulated glass panes. In the grand scheme of things, going cheap on these items can be detrimental to a home and its residents, and using superior products and incorporating the healthy home concepts outlined in this book will greatly reduce the need for some passive building practices if your property or budget constraints don't allow for them.

As a vital component of the home design process, you should hire a licensed structural engineer to ensure the foundation and framing are properly designed. Some states require home plans to be reviewed by additional teams of engineers for HVAC, plumbing, and electrical compliance. Make sure the engineering firm signs off on and stamps the construction plans. This means a professional engineer has designed the structural components to meet load-bearing requirements, which offers not only peace of mind for the homeowners but also important liability protection for the builder and all parties involved in the building process.

Much of designing a healthy living space is based on personal style, individual needs, and budget, so don't ignore any part of the design-build process. Before starting construction, make sure that your design allows you and your builder to incorporate all the necessary building material requirements and methodologies to make your home safe, sound, and chemical free.

A WORD TO THE WISE

We highly recommend you read this book in its entirety before embarking on your construction project.

Before finalizing any plans, you must make sure that all the members of your build team (which could include a builder, an architect or drafting team, a general contractor—or a combination of all three) are on board with your goals. And you need to outline the details well in advance.

We cannot overstate the importance of specifying which healthy building materials and methods you want to use up front. Homeowners should have an open dialogue with their builders about their product preferences during the design stage—well before breaking ground.

If your builder is not familiar with healthy building products and practices, your price per square foot may increase because they are entering unknown territory. Often, builders and subcontractors increase their prices to cover unknowns during a learning curve, such as unforeseen labor expenses and costly mistakes.

No one likes price tag surprises—this goes for DIY projects as well as commercial and residential jobs. Setting clear expectations before you start will help both you and your builder stay sane while keeping costs down.

key points

▸ Find a builder you trust (and make sure you're on the same page) before choosing your location.

▸ Plot your property in a way that serves you now and in the future.

▸ Open floor plans are ideal for most people and boost resale value.

▸ Position working and sleeping spaces away from sources of noise.

▸ Think about your needs for home offices, exercise spaces, parking, security, and more, both inside and outside the home.

▸ Be aware of the way sun, wind, and rain hit your house and plan accordingly.

▸ Hire a licensed structural engineer to ensure your foundation and framing are properly engineered.

▸ Your project will go much smoother if everyone is upfront and can agree on project expectations.

Ready to start building? Not so fast! Read about breaking ground before doing so in the next chapter.

Now that you understand the fundamentals of healthy building—including the "why" and "how" behind factors like form, function, and composition—let's talk building.

This section of the book outlines each step of this confusing and sometimes challenging process, from laying the foundation to constructing the walls and roof. Best to start at the beginning—when you break ground on your new home.

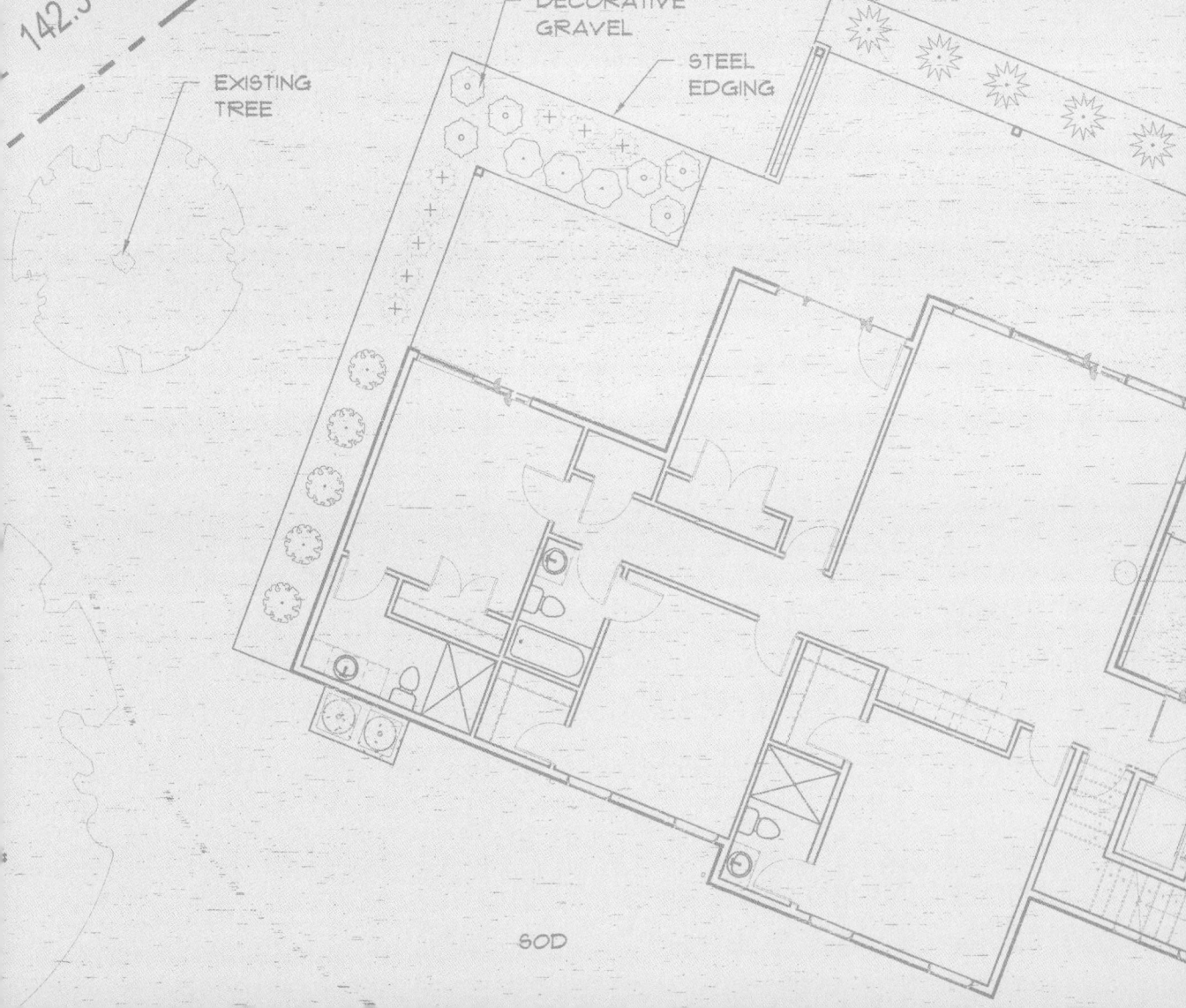

LIMITS OF SOD

SOD

DECORATIVE GRAVEL

STEEL EDGING

142'3"

EXISTING TREE

SOD

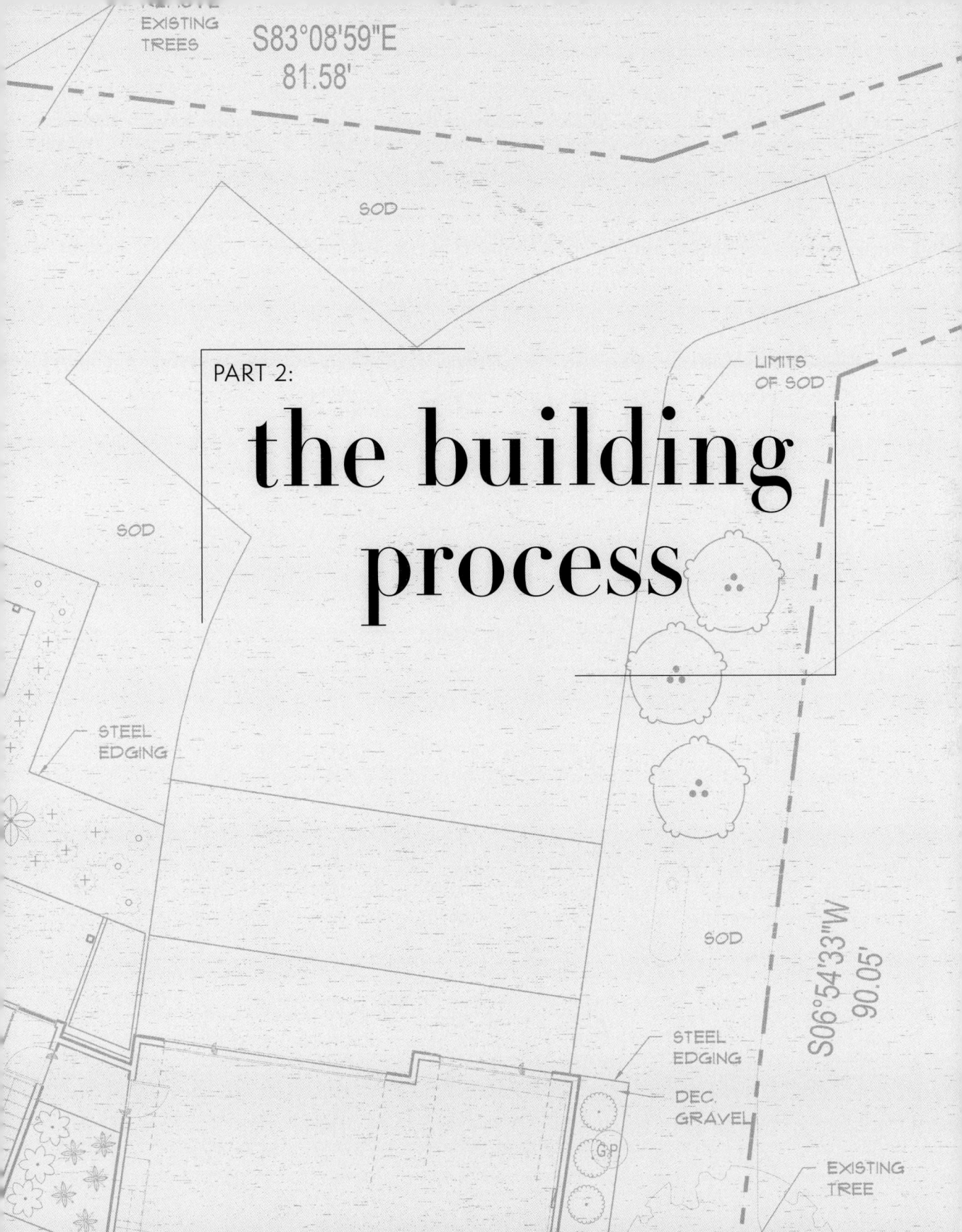

PART 2:

the building process

breaking ground

Breaking ground essentially means starting the building process. Dirt is moved, and excavation of the property carves out the footprint for your home. This is when things get "real" and you can begin to visualize your home. Breaking ground is truly groundbreaking in every sense of the word.

When working with construction drawings during the design phase, you should take soil samples from various parts of your property to determine your home's foundation and engineering requirements. Engineering a home on a rocky incline is a whole different ballgame compared to a flat piece of land with sandy soil. Topographical features such as sand, rocks, trees, water tables, and grade changes will all affect how your home needs to be anchored in place.

Soil variations can be surprisingly drastic at different depths, which is why it's important for your geotechnical engineer to dig to the correct depth and ensure you will have sound piers and footings (components of the foundation that support a building's weight). Soil conditions and grade often dictate the type of foundation needed, whether it's a slab-on-grade foundation or a pier-and-beam one. (Read more about those types in the next chapter.)

Your region and climate have a lot to do with soil considerations, too. Home builders in the Houston area, for example, must account for slimy layers of jumbo clay, which can lead to "slippage" or foundation movement if not engineered properly. Homes in the Dallas area, on the other hand, may experience dirt expansion and contraction during the hot, arid summer months. Dry soil that pulls away from a concrete foundation tends to crack, and critters love to make homes in these subsurface nooks and crannies. And then there are colder climates, where freezing and thawing conditions in the soil must be taken into consideration. Your structural engineer will dictate what type of building pad and constituents are best for your area's unique conditions.

HOW SOIL AFFECTS A FOUNDATION

Which soil is best for your home? The most important factors are stability and drainage. Here's a breakdown of soil types and how they affect a foundation.

CLAY

 Clay is composed of tiny particles that expand dramatically when wet, making it pliable and easy to manipulate.

 Being dry has the opposite effect; clay shrinks, grows hard, and can crack.

 These changes put pressure on a foundation, meaning clay should only be considered in dry climates not prone to flooding.

ROCK

 Limestone, sandstone, and bedrock have high weight-bearing capacities.

 The flatter it is, the better it is for building on. If it is not level, anchors may be needed.

 Rock makes for a decent foundation because of its stability and depth.

SILT

 Silt is also made of small particles that retain water well and lower the ground temperature.

 Expands and contracts due to moisture in the same way clay does, potentially weakening a foundation.

 Poor drainage qualities can make this type of soil unsuitable for building.

SAND/GRAVEL

 When sand and gravel are mixed and compacted, they make for a solid foundation.

 Together, these consist of large particles that do not retain much water.

 Note: As water drains away, the particles can loosen, eroding the foundation.

PEAT

 Peat is dark brown and consists of decomposed organic matter that is very porous.

 Peat shifts and changes, meaning it doesn't provide much stability.

 Peat is easily compressible, retains water, and dries brittle, making for poor support.

LOAM

 Most often, loam combines silt, clay, and sand. It is considered one of the best soil types for foundations.

 Loam has a dark color and feels dry and crumbly but doesn't change size or shape easily, even with moisture, making it very stable.

 The only real concern is other types of soil finding their way into the loam.

HIDDEN CHALLENGES

Sometimes, shallow aquifers and gas pockets present issues as well. It's always best to know where these exist on the property rather than finding them after you start to dig. The same considerations go for rocky soil—digging a hole in rock is much tougher than in soft dirt. Soil samples should be taken from various locations on the property and/or several corners in the slated building footprint. This process includes taking core dirt samples from varying depths to know exactly how deep each layer of material is. Think of it like different layers of colored sand in a jar.

Another possible challenge is land contamination, which requires the unique kind of soil test discussed in Chapter 3. For example, if your property used to be a gas station, conventional farmland, or a business that used industrial chemicals and/or solvents, we recommend a soil contamination test.

We recently worked with a client who was interested in investing in a property in an industrial district of a major US city. We did some research and found that the property backed up against a federally designated EPA Superfund contamination site. The investor promptly hired an environmental engineering firm to perform soil testing, and thankfully, the property was found to be free and clear of contamination. This is a good example of due diligence, as hiring a third party to do research up front when considering investing in a questionable property can save a lot of money later in the cleanup process.

Depending on the soil composition and your geographic location, radon may be an issue worthy of careful consideration as well. Radon is a colorless, odorless, tasteless radioactive gas given off by certain types of rocks and soil and is one of the leading causes of lung cancer. We generally incorporate a subsurface radon mitigation system under any home built on rocky soil or in areas with granite outcroppings. Strangely, a home with a radon problem may sit next door to a home with no radon issue whatsoever; levels can vary even within the same neighborhood. We err on the side of caution and install either passive or active systems depending on the situation.

The EPA recommends keeping radon levels in homes below four picocuries per liter of air. Interestingly, this level is still associated with a 1-in-100 lifetime risk for developing lung cancer for nonsmokers, and a 1-in-10 chance for those who smoke.[1] Those aren't great odds, so we don't mess around with this stuff.

GEEK BOX | ## the threat of radon

Radon is a gas that forms when naturally occurring radioactive elements, such as uranium, decay underground. This gas can escape into air and water, meaning it can exist indoors and outdoors. It is normally found in very low levels outdoors but can build up inside houses and other buildings, entering through cracks in the floors and walls or gaps in the foundation. Radon is colorless, odorless, and tasteless and is a leading cause of cancer, which is why a mitigation system is so important.

Check the map below to see if you live near a radon hot spot.

EPA Map of Radon Zones

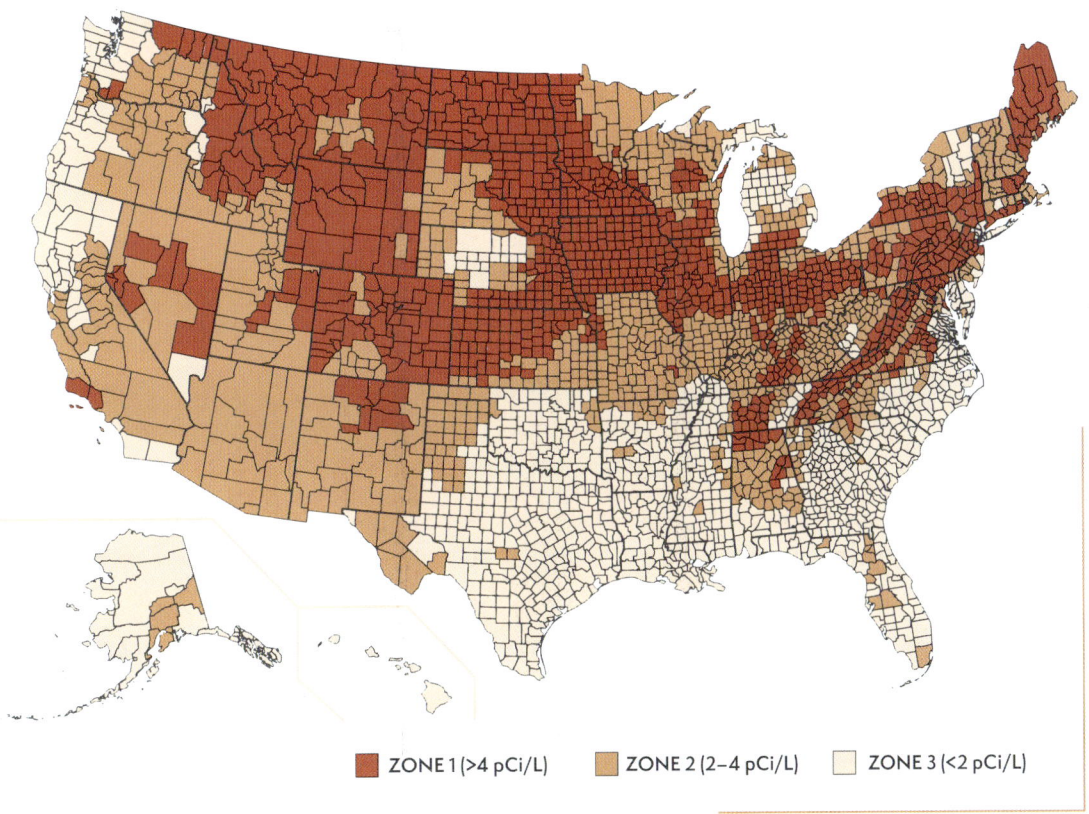

ZONE 1 (>4 pCi/L)　　ZONE 2 (2–4 pCi/L)　　ZONE 3 (<2 pCi/L)

SHEDDING WATER

To properly shed water from a foundation, you want to place the home on the highest point of the property. Topographical surveys come in handy when determining where that point is located. You should also choose a flat area if possible, because it means less concrete and soil will be needed to build up the lowest side of the house.

Our personal healthy home is built on the side of a large hill. Engineering, setting up drainage for, and backfilling a foundation on a slope can be substantially more expensive than building on a flat grade, and we ended up spending much more than we anticipated. In hindsight, it might have been better financially to position the house on a flatter part of our property; however, doing so would have meant sacrificing our lake view. Building is all about choices—giving here and taking there.

Our hillside healthy home. Being close to the lake provides a cooling breeze in the summer months and air rich with negative ions, which helps purify the air and decrease stress. [2]

Consider neighboring properties as well. Water runoff can present an issue when building downhill from another house. Proper drainage engineering is crucial for controlling the runoff on the side of a hill or mountain. Your yard and home may flood if the proper precautions are not put in place at the beginning of construction. Make sure your excavators properly grade the area around your home to shed water away from the foundation. Seeking guidance from a licensed civil engineer is highly recommended when devising your drainage plan on a tricky property.

We use large French drain systems across the front of a home's foundation wherever water running downhill might present a challenge. This design redirects water into a channel containing loose gravel and a perforated pipe to allow for easy draining without taking up unnecessary space. We recently installed a six-foot-wide French drain across the entire front side of a home; it ran underground and terminated at the downhill property line.

the advantages of French drains

Eliminate moisture: French drains let you capture and direct water from a higher spot to a lower one so you can keep your home or land from flooding.

Keep your yard pretty: French drains are located underground and range in size from a few inches to dozens of feet deep. The series of pipes captures water as it percolates through rock, meaning you can prevent ponding on the surface of drainage ditches without any visible pipes in the yard.

Are neighbor-friendly: French drains efficiently remove water from one location and safely transport it to another for controlled drainage. This way you're not unintentionally rerouting water onto your neighbor's property, which is illegal in some states.

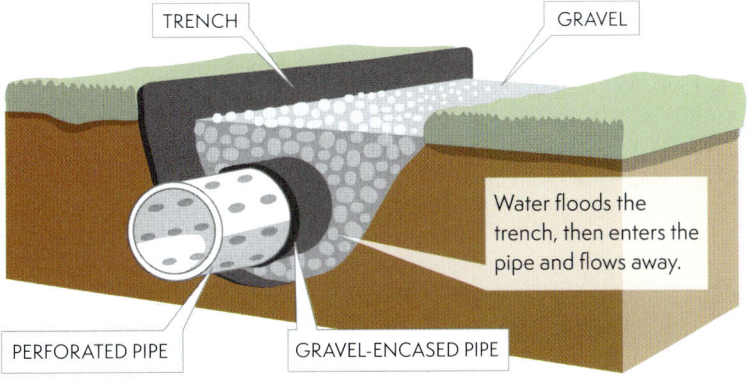

TRENCH · GRAVEL · Water floods the trench, then enters the pipe and flows away. · PERFORATED PIPE · GRAVEL-ENCASED PIPE

If you're building in a new development, make sure the developer has taken water runoff into account. We encountered a situation recently where a developer decided to add a road uphill from our client's new home, which was still under construction. The new road caused a massive influx of water to run directly into the home's front yard. It took quite a bit of communication to get this runoff issue resolved, but persistence pays off. If we'd chosen not to intervene and simply assumed that the developer had the drainage issue under control, the home would have had massive flooding.

SEE THE FOREST FOR THE TREES

A tree survey is recommended. Certain trees, depending on your region, can be considered nuisances. For example, some studies have found that mature cedar trees are water hogs and can consume thirty-three gallons of water a day from the soil. Removing these problem trees will help the beneficial ones flourish. Trees close to a home can also cause unintended problems with falling branches or clogged gutters from fallen leaves.

WHAT WILL A TREE SURVEY TELL ME?

A tree survey done by a land surveyor can provide a variety of useful information about the trees on your property so that you can make an informed decision about how to build your home. This info includes

- The species of the trees and the circumference of the trunks
- Physical measurements of the trees' driplines so that you can properly plot your land

Depending on the species and condition, you may want to remove problematic trees, such as those that appear diseased or unstable, and make plans to replace them with beneficial trees at the end of construction.

Trees and Shrubs Too Close to House

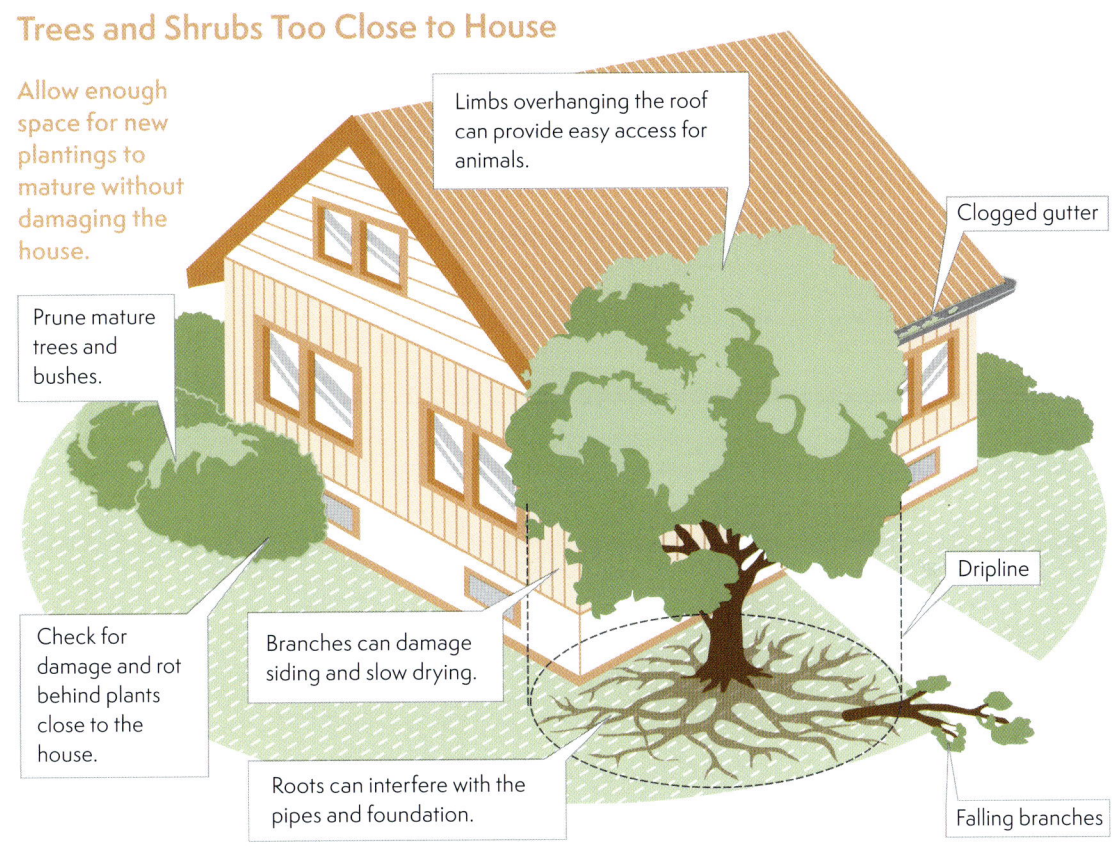

Allow enough space for new plantings to mature without damaging the house.

Limbs overhanging the roof can provide easy access for animals.

Clogged gutter

Prune mature trees and bushes.

Check for damage and rot behind plants close to the house.

Branches can damage siding and slow drying.

Roots can interfere with the pipes and foundation.

Dripline

Falling branches

A tree survey can help you decide where on the property to place your home. As a rule of thumb, a tree's dripline—or the area under the outer circumference of the branches where water will fall when the canopy gets wet—is approximately the same distance to which the roots extend. Depending on the species, some trees can root way past their driplines. The key is to keep trees far enough away from the home to prevent future root damage to the foundation or exterior.

Families with allergies might also consider removing trees that create a lot of pollen. Remember, you can always replace a problem tree with a healthy, beautiful one.

Another excavation tip is to make sure your utility lines and any abandoned underground pipes are accounted for. We ran into issues on a property in Houston years ago where an old, abandoned oil pipeline ran directly under the middle of the lot we were getting ready to build on. We commissioned a specialist to remove the pipeline safely so as not to contaminate the soil.

FINDING A SPECIALIST

Finding a trustworthy specialist involves many of the same steps as finding a good builder or GC. First, let the issue you need help with guide you toward a certain type of specialist. Say you need a civil engineer, for example (someone who can help provide information about site planning, surveys, erosion control, drainage, and establishing grades on the property). Next, check for appropriate credentials and licensing, which will differ depending on the area of expertise. Look into their local reputation, ask for a quote, and feel free to shop around. Most important, ask questions! Make sure your concerns are aligned with their priorities before signing anything.

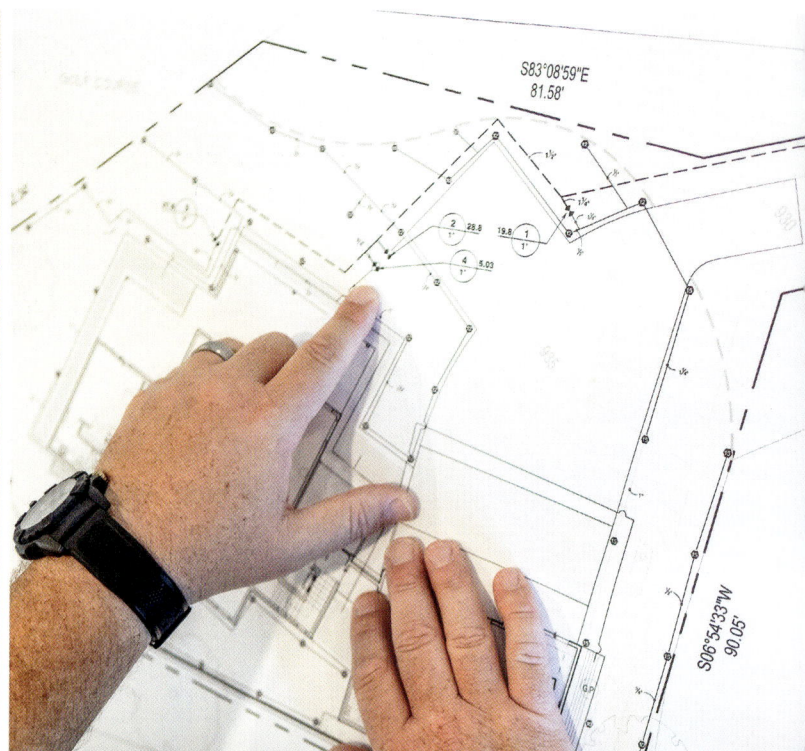

WHEN IN DOUBT, GOOGLE IT

Finally, a Google Earth search on your property may provide insights that are not available from a ground survey. Google Earth's aerial views can be especially helpful in evaluating neighboring properties and your surroundings. Recently, a Google Earth image revealed something unusual behind a thicket across the street from a homeowner's recently purchased property: a series of huge oil rigs.

Another example was a home that backed up to a dilapidated sewage lift station that needed repairs. Thankfully, we were able to work with the city to ensure the problem was fixed before the new homeowners moved in.

Taking the extra time to do your due diligence will lay the proper groundwork for a high-quality, structurally sound, and healthy custom home.

key points

▸ Take soil samples early to determine your home's foundation and engineering requirements.

▸ Be aware of hidden aquifers, gas pockets, and sources of radon.

▸ Conduct topographical, tree, and other professional surveys to get to know your land before you place and build your home.

▸ Ensure appropriate drainage by placing your house on the highest, flattest part of your property.

▸ Hire a civil engineer to address challenging drainage issues.

▸ Use Google Earth to get an aerial view of the area.

Want to know the pros and cons of different types of foundations? How about what a radon mitigation system does or how we keep moisture and critters at bay? We've got you covered in the next chapter, *Foundation*.

CHAPTER 6

foundation

Foundations are fundamental, which means they form the necessary base or core of support for any structure. The durability, longevity, and functionality of everything within your house depends on it having a solid foundation.

Careful planning is necessary during the design stage because foundations are both expensive to construct and difficult to alter once they're in place. Among the nuances to consider are flooring thickness and the locations of plumbing drains, floor outlets, and downdraft vents. All of these features are planned before the foundation is built.

TYPES OF FOUNDATIONS

While there are a number of foundation styles, these are the three most common types:

- **Pier and beam** uses buried concrete footings that anchor the foundation in the soil. Piers are then attached to the footing to support beams that connect directly to the joists beneath your flooring.

- **Basement** secures a house to a floor below ground level; although most foundations are sunk into the ground, only basements with foundations allow that space to serve as a finished room.

- **Slab on grade** uses a thick slab of concrete poured directly onto a layer of compacted sand or gravel (for drainage and cushioning) and steel rebar as reinforcement. Sometimes footings are needed depending on soil and terrain conditions.

For our purposes, we'll mostly be focusing on pier and beam and slab on grade. The major difference between these two types of foundations is that a pier-and-beam house is raised above the ground (typically with a crawlspace underneath), whereas a slab-on-grade house sits directly on the thick concrete pad.

Most homes built within the past few decades use slab-on-grade foundations unless there were specific geotechnical conditions that warranted a pier-and-beam system, such as layers of clay or expansive soils. Pier-and-beam foundations are also sometimes recommended for low-lying areas and coastal regions that are prone to flooding.

Both pier-and-beam and slab-on-grade foundations utilize piers that sometimes run deep into the ground. In colder climates, it's important to run the piers below the frost line to avoid underground damage from freezing and thawing conditions. These piers are structurally reinforced with steel rebar cages or metal columns, depending on the type of foundation, which anchor the home in place. At the bottom of the piers, footings disperse the tension from the structural load of the home.

To properly plan your foundation, you should hire a licensed structural engineer during the design phase. Only a licensed engineer can determine precisely how deep to go with the piers, how many piers to use, and where to place them.

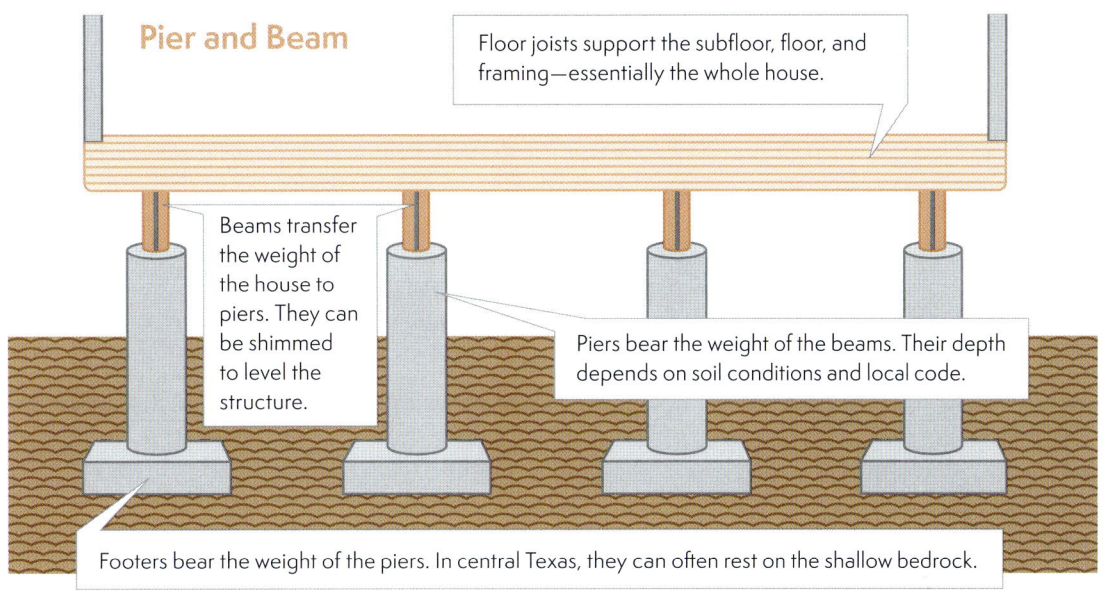

Pier and Beam

Floor joists support the subfloor, floor, and framing—essentially the whole house.

Beams transfer the weight of the house to piers. They can be shimmed to level the structure.

Piers bear the weight of the beams. Their depth depends on soil conditions and local code.

Footers bear the weight of the piers. In central Texas, they can often rest on the shallow bedrock.

Slab on Grade

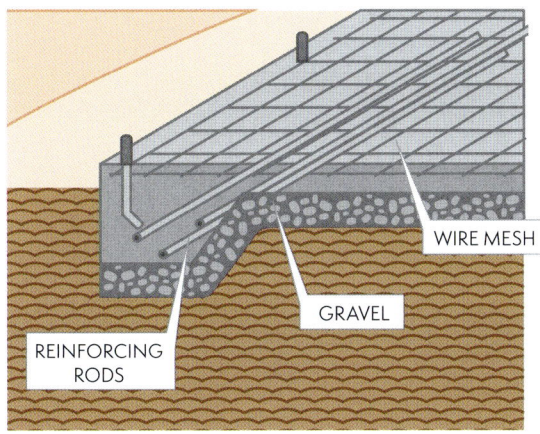

WIRE MESH

GRAVEL

REINFORCING RODS

Slab on Grade with Footings

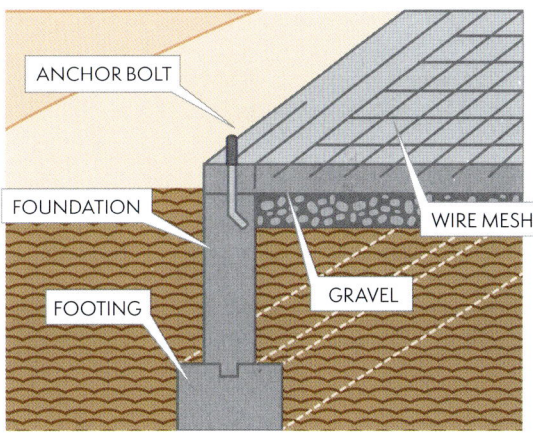

ANCHOR BOLT

FOUNDATION

WIRE MESH

GRAVEL

FOOTING

In regions where frost becomes an issue during the winter months, a structural engineer may suggest an added layer of subterranean insulation to prevent frost damage to the slab. A properly engineered foundation will address all requirements, such as depth, gravel size, and compaction methods.

FOUNDATIONAL PROS AND CONS

While there are some perks to building on pier and beam—such as cost savings from forgoing concrete and easier access to wiring and plumbing beneath the house—we do not recommend this type of construction unless soil conditions warrant it or a licensed engineer states that it is necessary. Although it has a few convenient features, we've seen mold issues time and again with crawlspaces. The dark, moist, stagnant environment provides the perfect breeding ground for fungus under your home—and a perfect place for unwelcome critters to sneak in. Snakes, varmints, and bugs all love crawlspaces, too.

Another challenge with pier-and-beam homes is ensuring that the home's envelope remains airtight. A leaky subfloor will pull air from the crawlspace directly into the living quarters every time there is negative air pressure within the envelope. This can happen frequently throughout the day, such as when you turn on the bathroom exhaust fan or the central air-conditioning system kicks on.

GEEK BOX

how do blower door tests work?

A blower door test is a tool used by energy auditors to determine a home's airtightness, so homeowners can reduce energy consumption, avoid moisture problems, and improve indoor air quality.

Blower door tests function by mounting a flexible panel with pressure gauges and a powerful fan to an exterior door. When active, it pulls air out of the house, lowering the internal air pressure, so that outside air flows in through any compromised openings.

It's important to hire a professional with calibrated blower door equipment, and there are several issues this test may help diagnose. It can

- Ensure your fireplace flue damper is functioning properly.

- Identify the need to caulk around windows and doors where seals need repair.

- Uncover issues with malfunctioning HVAC or exhaust fan dampers.

- Highlight areas where additional weather stripping should be applied around doors.

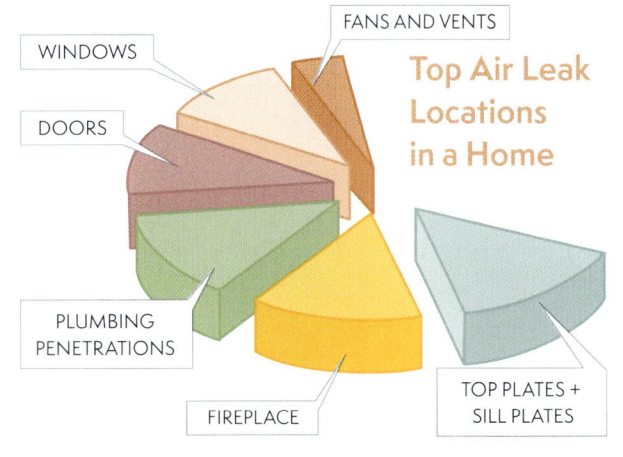

Top Air Leak Locations in a Home

FANS AND VENTS · WINDOWS · DOORS · PLUMBING PENETRATIONS · FIREPLACE · TOP PLATES + SILL PLATES

We recently worked with a homeowner in the Dallas area who was experiencing issues with indoor air quality and mold. For years he had worked with specialists to identify the source of the mold but had found no water leaks behind the walls. Even after the homeowner replaced his entire HVAC system, ran new duct work, and installed all new registers and plenums, the mold continued to creep back in, growing along the A/C vents and in the showers.

When JS2 Partners was called in to assess the problem, we noticed cracks in the walls due to probable structural issues in the foundation. It was an older pier-and-beam home, and after we performed a negative pressure test—which uses a tool that measures how airtight a house is—we found that a significant amount of air was entering from the subfloor. A quick inspection under the home confirmed a very dank and moldy crawlspace.

BIOBASED PRODUCTS are non-food commercial or industrial products composed of biological plant-based ingredients such as renewable agricultural or forestry materials. One major benefit of biobased products is that they are not petroleum based, which reduces harm related to toxicity and pollution.

THE PREFERRED FOUNDATION METHOD: SLAB ON GRADE

For the remainder of this chapter, we are going to focus on our preferred method of building a foundation: slab on grade. This method eliminates the issues with crawlspaces because there is no air pocket between the bottom of the home and the concrete pad. From both a structural and an indoor environmental standpoint, we believe that a slab-on-grade foundation is superior.

For slab-on-grade site preparation, make sure that your crew knows not to use petroleum-based dirt- and dust-control products. Site prep crews often use the same dust-control agents as road construction workers, but we advocate the use of biobased agents if dust control is required. We also recommend spraying the soil for termite prevention before installing the vapor barrier during the foundation process.

MITIGATING RADON

If you're building in an area with underground radon issues in the soil—or near granite outcroppings or with high natural levels of crushed granite in the ground—we recommend installing a radon mitigation system during foundation construction.

These systems consist of an interconnected series of perforated pipes laid out in a bed of gravel under the foundation before the concrete is poured. They lead to riser pipes that serve as vents for any radon gas that may accumulate under the foundation's vapor barrier. The risers run through the walls of the home and terminate at the roofline. Ensure these exhaust vents do not exit the home near windows or fresh air intakes. All riser pipes should be clearly labeled "radon vent pipe" to prevent a tradesperson from mistaking them for plumbing pipes.

A passive radon mitigation system allows radon to naturally vent up and out of the home's pipes safely, while an active system creates a vacuum using exhaust fans on the exterior of the home where the stack terminates. Keep in mind that if the mitigation system is an addition to an existing home, installation is a major project that requires drilling into the foundation, digging under the slab, and running a pipe up to the attic, so we highly recommend working with a professional.

For the layer of gravel, we prefer limestone or similar rock instead of crushed granite. Some granite has been found to be radioactive.

THE VAPOR BARRIER

It's best to go with a fairly thick foundation vapor barrier composed of high-quality polyethylene materials. Take care when laying and taping the foundation's vapor barrier. Seams should overlap by twelve inches and be sealed tightly with heavy-duty waterproof seam tape. Vapor barriers serve two purposes:

- They keep water vapor from entering your foundation.

- They keep pests and noxious soil gases out of your home.

GEEK BOX | is a vapor barrier alone enough?

In theory, an appropriately thick and properly sealed vapor barrier sheet should prevent gases and pests from entering a home. However, any microtears or degradations from heat that develop over time (which often occurs with inferior plastic products or improper installation) can create issues, allowing radon and insects to enter the home, along with other problems. If you live in a region where underground radon is common, visit the Healthier Homes website for details on how to install a radon mitigation system.

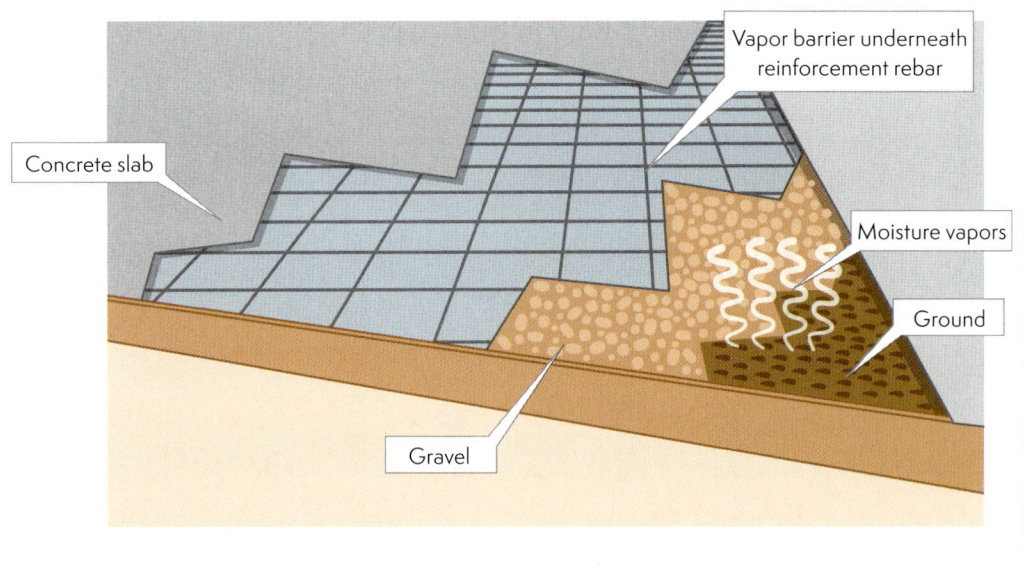

Vapor barrier underneath reinforcement rebar

Concrete slab

Moisture vapors

Ground

Gravel

Be aware of unforeseen ways in which chemicals may sneak into your foundation. Form-release agents are sometimes used on concrete molding materials to help with the removal process after the concrete has cured. (Forms are the cavities that are built to hold the concrete in place as it cures.) Opt for biobased form-release agents.

Metal rebar is an integral part of the foundation that reinforces a slab and gives the concrete its strength. It is common practice to tie rebar together at junction points with metal wire, as it substantially increases the structural integrity of the slab and helps evenly distribute tension. It's worth noting that some people have come to us with EMF concerns about the practice of tying rebar in foundations. We have not come across any information to support this concern and find the misinformation to be dangerous. Our recommendation is to follow standard rebar installation guidelines as specified by your licensed structural engineer.

INGREDIENTS FOR A GOOD FOUNDATION

Beyond the layers of gravel, vapor barrier, and rebar-reinforced concrete, we don't recommend placing much inside your foundation. This means not diluting the concrete with water before it is poured on-site. Aside from sleeves for floor plugs and data on the ground level, you should avoid running electrical wiring or the water main within the foundation if possible.

Placing water pipes in the foundation used to be a common practice, which unfortunately led to many moisture issue mysteries within homes. Slow leaks within the foundation are difficult to locate and may go unnoticed for years as they wick moisture across the concrete and up into the drywall and framing lumber. When wet, these porous building materials serve as the perfect all-you-can-eat buffet for a host of unwelcome guests, including mold, termites, and roaches.

Depending on the season and climate, your foundation subcontractor will know the best time of day to pour. Remember that the top of the home's slab should be at least six inches off the ground around the entire perimeter.

GEEK BOX

the surprising water-wicking ability of concrete

Many people—even those in construction—mistakenly think concrete is watertight. Concrete does a pretty good job of containing liquid, but its permeability depends on a variety of factors.

While laying the slab, you want it to be wet—without moisture, placing and strengthening can't happen. As this process occurs, the mixture forms tiny pathways called capillaries, which excess water uses to move up and out of the slab—if there's a proper vapor barrier beneath it. But these capillaries also provide a path for moisture to get into the concrete, especially during flooding or if the vapor barrier fails.

One of the biggest factors is the quality of the vapor barrier placed between the foundation and the ground. Underneath nearly every concrete slab is damp earth, which means there is always water vapor trying to get into the foundation. In fact, without an adequate vapor barrier, the slab may never dry out. This can result in significant structural damage and mold problems, so remember to incorporate waterproofing measures into the design and build process.

THE BEST CURE

A slow concrete cure is always best for maximum structural durability. If you're pouring a slab during the hot summer months, pouring early in the morning is a good idea. Also, connecting an oscillating sprinkler to a potable water source and setting it up over the foundation may help slow the curing time.

Avoid using chemical curing accelerators and similar additives, which often contain formaldehyde and other off-gassing materials. Concrete is porous and will continue to off-gas these chemicals into your home.

If you're building on a slope, drainage also needs to be considered during foundation planning. Properly rerouting water around a slab through soil grading and installing a large-scale French drainage system (if required) is much more efficient if planned for during the foundation stage. This is a good time to consider how you will tie in the gutter downspout system.

types of drainage systems

We talk a lot about French drains because they are among our favorite drainage systems, but there are several types, and many of them can work together.

Surface system: One of the most common residential drainage systems, a surface system uses channels or ditches to prevent water accumulation and flooding. This is a simple, efficient, reliable option for getting rid of surface water.

Subsurface system: This type of system is implemented beneath the top layer of soil to remove excess water at the level of roots. A large, perforated pipe wrapped in cheesecloth is typically placed underground to capture and route water away. Technically, a French drain is a subsurface system.

Slope system: This system is designed to let water flow down and away from a structure, using gravity to guide the water to an area drain or a drainage pipe. Creating natural swales or low-lying paths for water to follow also takes advantage of the natural topography of the land.

Downspouts and gutters: As a structure's first line of defense against stormwater, downspouts and gutter systems route water to other drainage systems outside the house, often into a buried drainpipe or rain catchment system.

THE PROBLEMS WITH BASEMENTS

Waterproofing membranes are critical if a home will be built in a subterranean area. For example, a home with a walk-out basement (sometimes referred to as a daylight basement) that is built into the side of a hill requires adequate waterproofing on the exterior of the subterranean retaining walls. We double up by rolling a layer of waterproofing material on the exterior of the retaining walls and then adhering an additional membrane of thick waterproof rubber. This is only one piece of a complex puzzle for these types of homes, which have a highly involved build-out process with extensive details from the structural engineer.

Below-ground basements are popular in some regions. Although basements provide storage space and offer a safe haven in tornadic weather, they also come with humidity and moisture issues. For this reason, we advise homeowners to forgo basements on new builds if possible.

key points

‣ Your foundation is foundational. Do the necessary planning and have it properly engineered to ensure the durability, longevity, and functionality of your home.

‣ Be aware of the pros and cons of pier-and-beam versus slab-on-grade foundations (hint: we prefer the latter, if doable).

‣ This is the time to plan for and install proper drainage as well as an active or passive radon mitigation system if warranted.

‣ Unless you have an unavoidable need for floor plugs, don't place wiring or water mains inside your concrete slab foundation.

‣ Basements tend to have moisture issues. We don't recommend building one.

It's time to bring your designs to life in the next chapter, *Structure.*

structure

Now that your foundation is in place, it's time to bring your home to life. This is when things truly start to come together and the house begins to take shape vertically.

First on the list should be to check local building and fire codes. They will help shape this part of the build and are important to implement correctly the first time. (If you are working with a reputable builder, they should handle this task for you.) Neighborhoods often have architectural committee (ARC) guidelines, which also must be taken into consideration.

BUILDING STRONG BONES

The structural components are the first major pieces to be erected. Think of the structure as the bones of the home.

Builders use several types of materials to frame a house. Most residential projects use lumber for framing the walls, beams, joists, and roof decking. Depending on the region in which you live, different species of readily available wood may come from the spruce, pine, and fir families. In Texas, we generally use pine.

A framing package for a house is a big-ticket item. Even with recent increases in the price of lumber due to tariffs and high demand, wood is still a more economical option for framing than alternatives such as steel. We have found that the entire assembly process goes quicker with lumber, simply because welding is not necessary and the experienced labor pool is larger. Plus, trees are a renewable resource that is readily available in most regions. Keep in mind that structural engineers may call for some steel in the framing process for areas that need reinforcement.

The Bones of a House

STUD: A vertical piece of lumber that supports the walls

GIRDER: A strong piece that transfers a load to its supporting wall

BRACE: A vertical or diagonal piece used to strengthen studs

JOIST: A level piece that supports the floor or ceiling

STRUT: A piece between two studs that increases stability and strength

RAFTER: A sloped piece that supports the roof

TIE BEAM: The beam against which the rafters rest

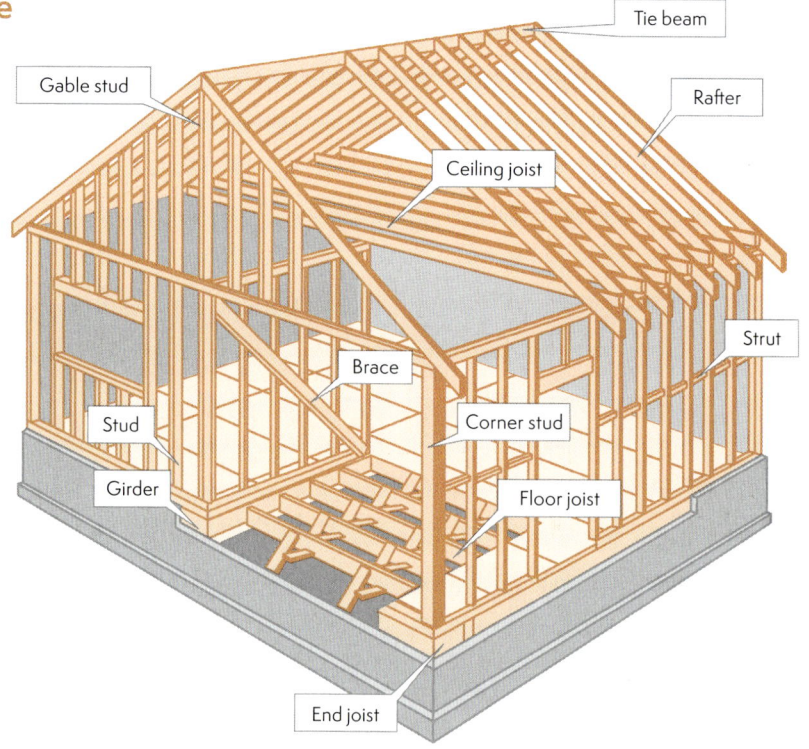

Tie beam

Gable stud

Rafter

Ceiling joist

Strut

Brace

Stud

Corner stud

Girder

Floor joist

End joist

types of lumber

The two types of wood are **hardwood** and **softwood**. Hickory, maple, and oak are commercially available examples of hardwoods—expensive, but good for cabinetry, flooring, and woodworking. Softwoods include cedar, pine, and fir, which are typically more economical and easier to source, making them ideal for home construction.

Light structural lumber: This timber milled from softwood trees is primarily used for residential construction. These are your standard-dimension boards that can be used as studs, joists, planks, beams, and so on.

Heavy timber: Any lumber over 4.5 inches wide that can bear heavy loads and ensure longer life spans. This timber is commonly used for post-and-beam constructions.

CONTAMINATED LUMBER

Before framing begins, it's important to ask your framing supplier if the lumber has been sitting in the yard for a while. Fresh framing packages are best to avoid mold growth on planks. We haven't had this problem lately because the construction industry has been very busy, but checking is still a good idea.

Quite a few homeowners are concerned about potential mold contamination on their lumber. In fact, we've heard of some mold "experts" recommending erecting a giant tent over a home during framing to supposedly reduce the risk of mold growth. While tenting an entire home during residential construction is possible, it can be incredibly costly and is rarely warranted. It can also be downright dangerous should a windstorm ensue. Plus, too much dust and heat trapped inside the tent could make for an unhealthy work environment.

(There are times when wrapping portions of a home is important. We have wrapped scaffolding along the exterior of a home during the removal of EIFS, a Styrofoam-based stucco product. More on that later.)

Another common request from homeowners is to cover the wood with a tarp while it's sitting on-site. The truth is that all lumber has been rained on many times before it ever reaches the job site. From the time it's cut down to its trip on a train or semitruck, chances are there was some rain along the way. A tarp is not advised for framing lumber because any moisture that blows under the edges or wicks up from the ground will remain trapped underneath, which could actually promote mold growth.

CLEANING MOLD

The objective in cleaning mold is to remove all fungal material—including the roots, spores, and spore-producing structures in the mold—from the surface of the wood. That does not necessarily mean removing stains from wood.

We outline our own industrial-strength (yet still healthy) method for cleaning up mold in Chapter 20, but you probably have some common antifungal cleaning agents around your house already. Reach for hydrogen peroxide, white vinegar, or good old-fashioned baking soda for your own mold spot-cleaning needs.

terpenes

Terpenes are fragrant oils found in softwood trees, weeds, and grasses that are classified as naturally occurring VOCs. Pinene is a popular terpene found in pine needles and pine wood that has a strong pine tree scent. It is also present in conifers, rosemary, basil, and even orange peels. Produced by plants as a natural protection against predators, terpenes are useful in small quantities in cleaning products and apothecary remedies. However, constant exposure to terpenes inside a home can irritate the sinuses and respiratory tract. Some people with allergies cannot tolerate terpene odors. This is why we advise against log cabin construction.

It's important for your construction team to focus on getting the framing, sheathing, and roofing in place in a timely fashion. The longer a home's interior lumber is exposed to wet weather, the greater the chances it will become contaminated.

If you live in a humid climate, investing in a moisture meter isn't a bad idea. There is no hard-and-fast number to shoot for, but you typically want lumber readings below 15 percent after the house is dried in—meaning after the building has been completed enough to keep out bad weather.

Borate-treated lumber, often referred to as "blue wood," is a popular option in Canada for keeping bugs and mold at bay during framing. Borate is a naturally derived substance that is not harmful to people or animals but acts as a biocide that protects wood from fungi, rot, and wood-loving insects like termites. Sadly, blue wood isn't readily available in the US, and the price is substantially higher than that of regular framing lumber. We do advocate for the use of borate-treated sill plates, which are easier to source.

All wood has the potential for mildew growth, so don't be surprised if your framing package is delivered with some moldy planks. We simply discard the wood that has visible mold as we work through the framing process. We've had homeowners ask us if they can be on-site to sort through the lumber packages and discard wood that appears dirty. Lumber is delivered by the truckload, so accommodating this request simply isn't always feasible.

GEEK BOX | **dangerous lumber**

Although wood is a natural product, that doesn't always mean it's safe. It can off-gas substances both man-made and natural and present other threats to a healthy home. Here are just a few things to watch out for with lumber:

· **Terpenes:** As discussed earlier, these compounds found in most plants create their distinct smells but are still considered VOCs and can affect humans in unpredictable ways, such as by causing allergic reactions. Hardwoods tend to have fewer terpenes than softwoods.

· **Fumigation:** A variety of insecticides may be used to keep bugs away from wood. Avoid wood treated with insecticide.

· **Formaldehyde:** This is often present in engineered woods, but all lumber naturally emits some formaldehyde VOCs.

· **Moisture:** Freshly cut wood contains moisture. At some point, it will achieve "equilibrium moisture content," which means it reaches a balance with the atmosphere and stops absorbing or releasing moisture. Wood should be allowed to reach equilibrium before additional materials or finishes are applied.

ARE STEEL FRAMES WORTH IT?

Steel framing is often used in commercial building and is sometimes necessary for structural purposes, especially if a building is over four stories tall. Some homeowners have expressed interest in using metal framing to avoid mold growth on the wood if a leak were to happen after construction, but we generally advise against it on residential projects.

Metal frames within a home's walls, ceilings, and floors can reflect and carry electrical frequencies and compound their physiological effects on the occupants. Dr. William Rea, a well-respected surgeon and one of the founding fathers of environmental medicine, described a group of mold-sensitive people who moved to New Mexico in search of drier, cleaner air. They built their homes using metal frames to avoid lumber but began to feel the amplified effects of EMFs almost immediately. Grounding equipment did not make much difference. Since we strive to build our homes with as little EMF pollution as possible, we have taken his advice to steer clear of all-metal frames.

Fabricated metal materials such as cold-formed steel also tend to have a thin film of petroleum residue from the manufacturing process. If you decide to go with metal framing, these residues should be washed off before installation. Industrial-strength nontoxic detergent mixed with warm water in a pressure washer will work to remove the residue.

For these reasons—and to keep costs down—we generally stick with timber for residential framing.

FRAMING METHODS

We prefer platform framing for smaller structures such as houses. With a platform frame, the first-floor wall sits on the foundation and each subsequent wall sits on subflooring, so multistory houses are built one level at a time. Each floor provides a platform that supports the next series of walls.

The foundation, whether it's pier and beam or slab on grade, provides the first platform on which the house is built. Each platform's frame must adhere to regional building and fire codes. Dimensions for window heights, door heights, and ceiling and hallway transitions are all accounted for during this phase. Studs, joists, and rafters provide the stability to support the structure and roof deck overhead. Although this part may seem rather technical, sometimes it's easier for a homeowner to communicate with their builder if they have an understanding of basic building process concepts.

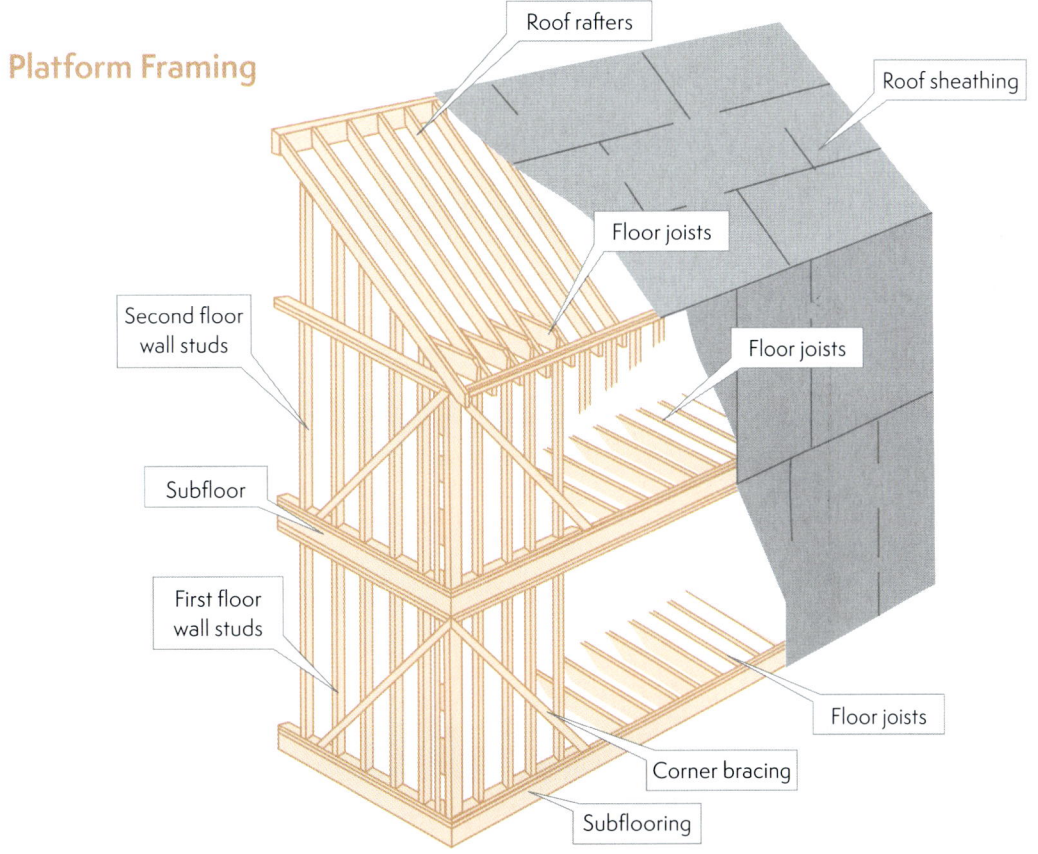

Platform Framing

Roof rafters

Roof sheathing

Floor joists

Floor joists

Second floor wall studs

Subfloor

First floor wall studs

Floor joists

Corner bracing

Subflooring

ALTERNATIVE BUILDING SYSTEMS

Several homeowners have approached us about using alternative building systems to frame up a home, such as the cement building block systems employed by companies that claim to use eco-friendly materials in their processes. These are variations of insulated concrete form (ICF) construction, which basically creates a concrete wall around the home.

The appeal is that these materials are touted as sustainable and natural. However, after conducting an extensive review of the building processes, methods, and materials that are used in many of these systems on the market, we have concluded that current iterations of ICF construction do not offer a way to build a healthy home that meets our standards.

The raw materials used in these systems are questionable in terms of meeting not only strict energy codes and local building requirements but also our accepted levels of nontoxicity. Furthermore, we have EMF concerns about the caged rebar within the walls, as well as concerns about any process that involves filling the walls with a cement mixture that cures solid, which could make it difficult to perform maintenance, repairs, and renovations down the road. Plus, ICF is not a vapor barrier, which means the indoor environment of an ICF home cannot be controlled. Thus, we do not use ICF construction on any of our healthy homes.

key points

▶ Familiarize yourself with local building and fire codes before starting the framing process.

▶ Learn about the types of lumber available in your region.

▶ Before you start building, take steps to avoid using contaminated lumber.

▶ Don't put your lumber under tarps. A little rain won't hurt it.

▶ Avoid steel frames if possible. They can reflect EMFs and may be coated in petroleum residue.

▶ Use caution when considering alternative building methods like ICF construction.

Now that you're confident your home's bones are strong, let's move on to the exterior construction portion of the process. In the next chapter, *Roofing*, you'll find out about the best methods and materials for covering your home.

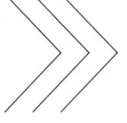

roofing

If a home's framing is its bones, then the sheathing and roof deck are its skin. And, just as our skin protects us from the elements, a structurally sound exterior will withstand the harsh outdoor environment for decades to come. That includes wind, rain, hail, extreme temperatures, UV rays, and pollution.

In fact, exterior construction has multiple layers, just like our skin. A roof is the topmost layer, and believe us when we say that a strong, waterproof roof can make or break a healthy home. As such, we spend a great deal of time on roofing, during both the design phase and the construction process.

THREE RULES FOR ROOF DESIGN

When it comes to roof design, we follow three cardinal rules:

1. **No flat roofs.** As much as we love the look of modern homes, there's always a way to design a home without a totally flat roof. A good rule of thumb is to design a roof pitch to at least $\frac{1}{12}$ (or 8 percent). This pitch ensures water will run off the surface and not pool on top.

2. **Consider the watershed direction.** We design our homes with a roofing system that ensures water flows away from the home, not toward it. But we often see this done improperly, which can lead to water intrusion issues down the road.

3. **A roof must protect the sides of the home, too.** The longer the overhang, the better the roof can keep rain and ice off the siding and windows.

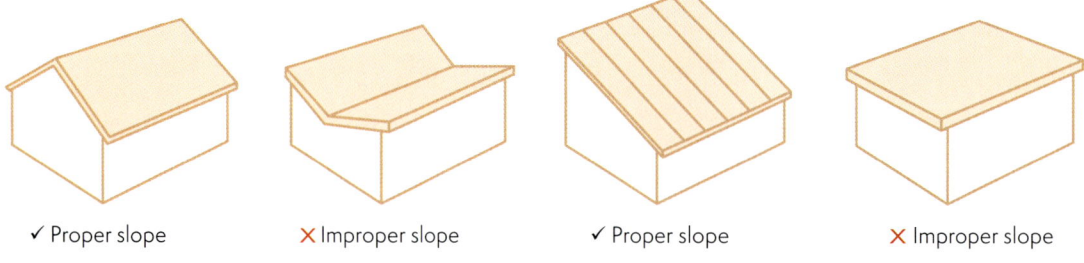

✓ Proper slope   ✗ Improper slope   ✓ Proper slope   ✗ Improper slope

In recent years, some modern homes have been designed with inward-angled roofs. As a result, the roof drainage systems are installed in central locations on the roof with pipes carrying water through the interior walls and out at the foundation. This is not only unnecessary but also a recipe for disaster should the drains on the roof become clogged. More holes in the roof mean more opportunities for rainwater to enter the home. In this example, multiple factors compound the threat of water intrusion:

- Even with adequate flashing—a thin material such as sheet metal used to seal joints in a roof—rain can produce a surprising amount of force.

- Leaves and debris can block drains, causing water to pool on the roof. Standing water is heavy and often corrosive, meaning it can breach waterproofing materials.

- Drain lines that exit through the foundation also present an opportunity for pests or varmints to move in and make themselves at home within your walls.

- Gutter line exits should be routed away from the home's foundation so that water does not pool around the slab. In the example above, the interior gutter system requires external drainage pipes to safely carry water away from the house—likely buried underground, which means a cracked pipe could go undetected for a very long time.

- A flat roof requires homeowners to be on their exterior maintenance game, which often includes the periodic application of hot asphalt. We avoid asphalt-based products because they have tremendous off-gassing potential, especially in the hot summer months.

How Flashing Protects a Home

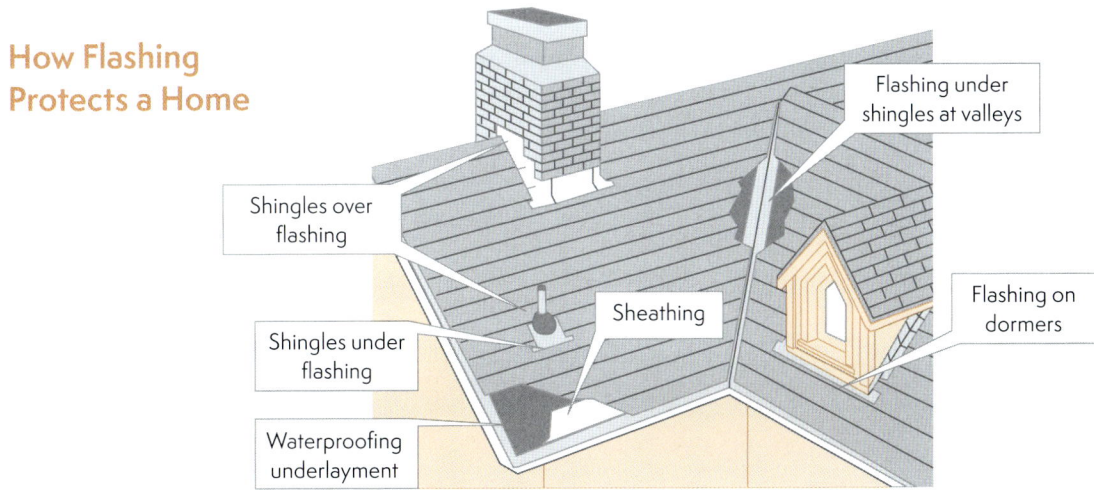

Flashing under shingles at valleys

Shingles over flashing

Sheathing

Flashing on dormers

Shingles under flashing

Waterproofing underlayment

GEEK BOX
the force of water

Aside from your home's location and the construction materials used, the amount of damage rain can cause depends primarily on volume and duration. For rainfall to qualify as heavy, it must exceed 0.3 inch per hour. A storm that deposits a half-inch of rain can hit a residential roof with hundreds of gallons.

Leakage: Heavy rain will reveal any weak spots in your home, entering through cracks in the doors and windows and any damaged parts of the roof such as split shingles or gaps around vents or chimneys. Leakage into the attic can ruin insulation, damage structural components, and cause mold problems.

Flooding: Repeated downpours can saturate the soil around your house, which can affect the foundation or basement. Hydrostatic pressure can push water from the soil through cracks and gaps in concrete, creating other moisture-related issues.

Exterior damage: Clogged or improperly sized gutters can overflow and penetrate exterior siding, which is designed to repel occasional rainwater but not to endure continuous exposure. This kind of penetration can result in rot and mold.

Think of the roof deck as the innermost layer of protection for the top of the house. These large, thick sheets of plywood (or roof sheathing) need protection from the outdoor elements. A waterproofing membrane should be used on every roof. While some builders only apply the membrane in the valleys, we cover the entire roof deck to ensure additional protection against ice and water.

For the roof deck itself, numerous materials are available to fit all styles and budgets. Some are great while others are lacking in quality and durability.

ROOFING MATERIALS

Wood shingles and shakes date back to ancient times, but their use has declined over the past century due to their flammability, potential to stain, and shorter life span.

For the most part, composite roofing (also known as asphalt shingles) has replaced wood as an economical option with a decent track record of durability. However, roof leaks are not uncommon because composite roofing sheets must be nailed down. Plus, when asphalt grows hot during the summer months, it can release harmful VOCs into the air. For many, asphalt fumes can be quite nauseating (think of the smell of freshly paved roads).

Slate, cement, and tile are all excellent roofing options that will outlast composite by decades. No nails are required for installation, which is a huge bonus in terms of waterproofing protection, but these materials are quite a bit heavier, meaning the home will need to be structurally engineered to support the weight.

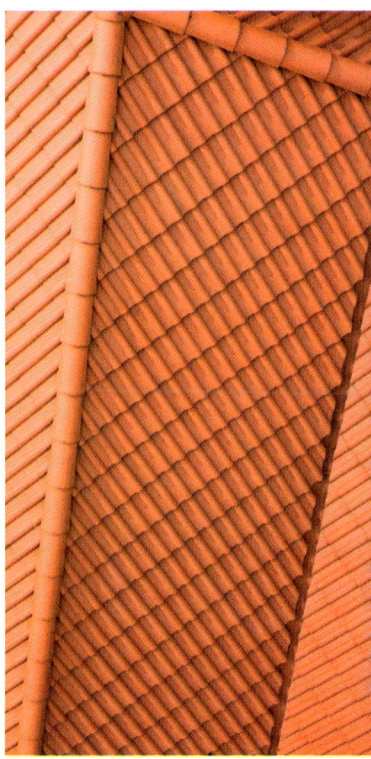

GEEK BOX the harmful effects of asphalt

Asphalt is traditionally made using coal tar and bitumen, a petroleum-derived product that serves as a sort of waterproof glue. Bitumen is also used in other construction materials such as flooring adhesives, insulation, rubberized building products, and more.

Unfortunately, both tar and bitumen are known to release high levels of VOCs that can cause the following health complications in humans:

- Fatigue
- Rash
- Nausea and vomiting
- Headache
- Eye and throat irritation
- Cancer
- Reduced appetite
- Dizziness
- Cough
- Shortness of breath

We do not build with coal tar or bitumen products in our healthy homes.

In the nineteenth century, metal roofing became popular because it could be manufactured in large sheets that were cheaper and lighter than alternatives. These metal roof systems were mainly used for commercial and agricultural buildings and required many screws to hold down the panels—the main problem with metal roofs of the time. Even when primed with silicone to improve water resistance, the screws still needed to be replaced every ten years or so, which made leaks a frequent problem.

The invention of standing seam metal roofs, however, was a game changer. No nails or screws are required for installation; instead, the panels are crimped together. The result is a waterproof roof that is lighter and more economical than slate, cement, or tile. Installation is quick, and the warranty from reputable manufacturers can't be beat.

That said, standing seam metal roofs have a few downsides. For example, they are more time-consuming to replace. Because of the crimping, if hail damages a single area on the roof, multiple layers must be removed to replace the damaged sheet. Enhanced EMF may also be an issue if the homeowner plans to use devices that emit radio and microwave frequencies in the home (mainly Wi-Fi and Bluetooth).

Of all the options, we believe the standing seam metal roof is the winner when it comes to waterproof protection and getting the best bang for your buck. If your budget allows for clay, stone, or cement tiles, those are also wonderful options for a healthy home.

FIREPLACES: YES OR NO?

Earlier we mentioned that fewer holes are better when it comes to roofing. The chimney is no exception. We often design homes without chimneys altogether. But many people value fireplaces as the focal point of the living room. They're perfect places to hang Christmas stockings and often represent warmth, comfort, and nostalgia.

Good news—we have an economical alternative for you! LED fireplaces can give the look and cozy feel of a real fireplace. Some can even change colors or provide warmth with built-in electric heaters. The only functional disadvantage of these faux fireplaces is that they are useless at providing warmth during a power outage.

Direct-Vent Fireplace

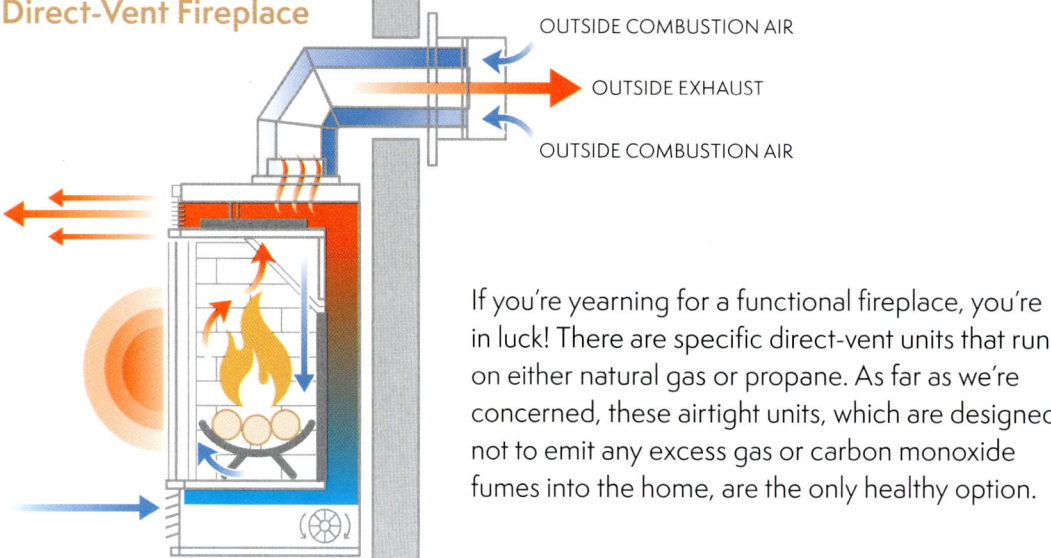

OUTSIDE COMBUSTION AIR

OUTSIDE EXHAUST

OUTSIDE COMBUSTION AIR

If you're yearning for a functional fireplace, you're in luck! There are specific direct-vent units that run on either natural gas or propane. As far as we're concerned, these airtight units, which are designed not to emit any excess gas or carbon monoxide fumes into the home, are the only healthy option.

Some types of woodburning stoves are excellent sources of heat and are designed to vent smoke directly out of the home. The drawbacks are that your home still may have a slight smoke smell, and they do produce smoke outside the home (which is pollution). However, they burn more efficiently than woodburning fireplaces.

When it comes to building a working fireplace, your material options are many. You can use stone, brick, concrete—the list goes on. And a fireplace has more components than you may realize, ranging from the foundation to the flue. Remember to consider maintenance and safety requirements, as the chimney will require periodic cleaning by a professional.

Parts of a Fireplace

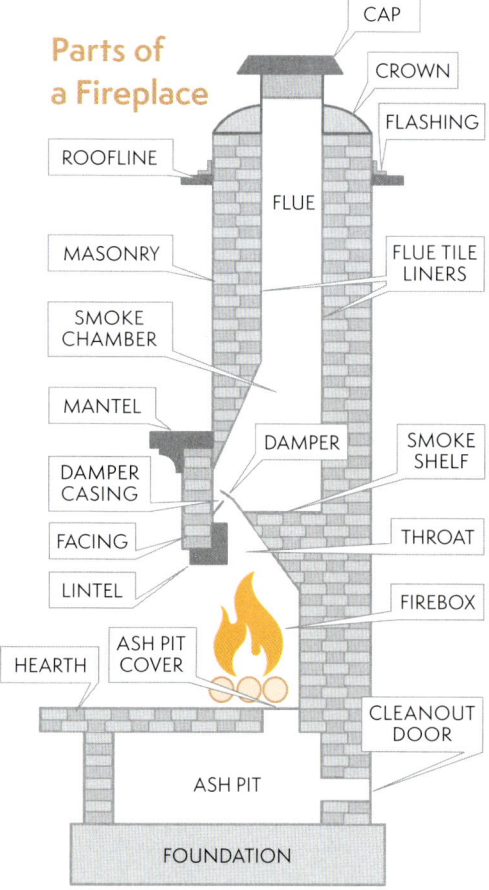

CAP

CROWN

FLASHING

ROOFLINE

FLUE

MASONRY

FLUE TILE LINERS

SMOKE CHAMBER

MANTEL

DAMPER

SMOKE SHELF

DAMPER CASING

FACING

THROAT

LINTEL

FIREBOX

ASH PIT COVER

HEARTH

CLEANOUT DOOR

ASH PIT

FOUNDATION

DEADLY GAS LEAKS: WHAT YOU SHOULD KNOW

Natural gas or propane can leak into the home due to loose fittings, pipe ruptures, or other issues. When this happens, the results can be extremely dangerous. Fumes can accumulate inside the home, leading to fires or explosions and health problems, including

- Ringing ears
- Reduced appetite
- Chest pains
- Nosebleeds
- Dizziness and light-headedness
- Drowsiness
- Flu symptoms
- Difficulty breathing
- Neurological damage
- Poisoning and death

You should also be aware that these gases can produce carbon monoxide as a by-product of incomplete combustion. Carbon monoxide poisoning can cause similar symptoms, even severe ones like loss of muscle control and death.

Carbon monoxide is odorless, and the distinctive sulfur scent of natural gas can fade, so you cannot rely on your sense of smell to detect a leak. Get a carbon monoxide detector and be aware of your symptoms when you leave home and come back.

If you suspect a mild leak, open your windows, turn your pilot lights off, and contact the gas company and/or fire department. If you have severe symptoms, go to the emergency room immediately or call 911.

A WORD ABOUT ATTICS

You may be wondering how all this affects your attic space. Since energy codes require homes to be built airtight, reputable builders are not installing the once-common vented soffits, which refer to the undersides of overhanging elements such as roof eaves. This style of home was designed to permit air to flow through tiny holes that allowed the attic to breathe and release heat up through ridge vents.

To meet today's strict energy codes and efficiency standards, we now build homes as enclosed building envelopes. No more cold attics in the winter and hot attics in the summer, which also allows for complete control over the conditioning of the indoor air.

The building envelope includes the roof, walls, and floor. Each part of this enclosure faces different challenges in providing structural stability and keeping air, water, and heat or cold from getting in while giving water a way out. Creating an airtight barrier to prevent air leakage can increase comfort while reducing energy bills and carbon emissions. But it can also trap harmful VOCs within the home if not constructed properly.

Many builders do not include the attic in the building envelope, but bringing the conditioned space into the attic allows for maximum control over indoor air quality and thermal comfort. Our enclosed building envelopes condition the attic, which we build as another nontoxic area suitable for storage and bonus living space. Who wouldn't want all the space under their roof to be clean and usable, including the attic?

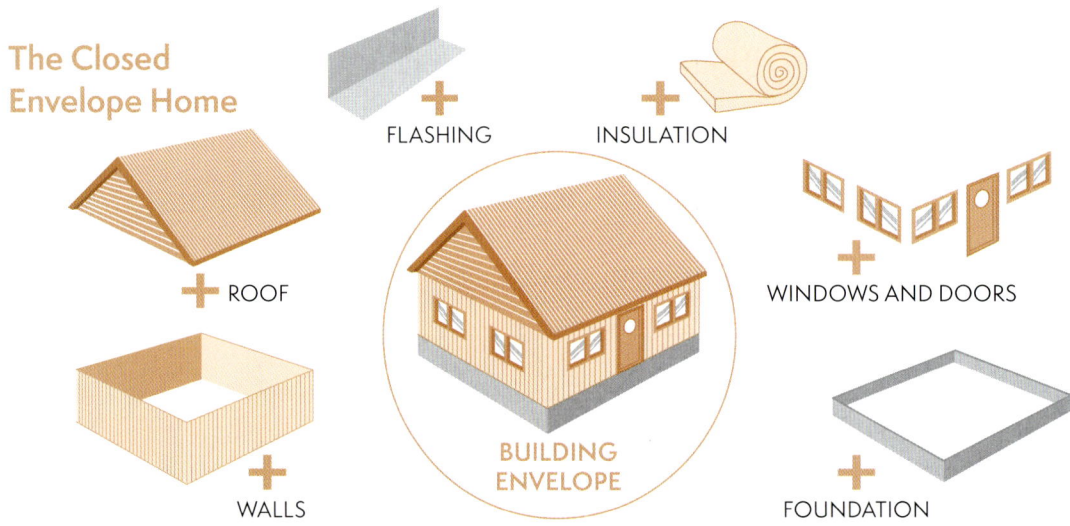

The Closed Envelope Home

FLASHING + INSULATION +

ROOF +

WINDOWS AND DOORS

BUILDING ENVELOPE

WALLS +

FOUNDATION +

key points

▶ A house's exterior is its skin, protecting it from wind, rain, hail, extreme temperatures, UV rays, and pollution.

▶ We have three rules for roof design: no flat roofs, consider watershed direction, and design to protect the sides of the house as well.

▶ Be aware of the force of water and the power of good drainage when considering roof designs, especially more modern ones.

▶ A good roof deck should always be fully covered with a waterproofing membrane.

▶ Standing seam metal roofs are great, as are clay, stone, or cement tile roofs.

▶ For waterproofing reasons, we recommend against traditional chimneys and fireplaces, but there are high-tech alternatives.

▶ We keep the attic inside the building envelope.

Now that you have the roof down (or up), perhaps you're wondering what to do for the sides of your house. Head to the next chapter, *Exterior*.

exterior

Now that we've covered the roof, let's move on to the other exterior items, starting with the sheathing, which covers the sides of the home. Like roof decking, sheathing serves a dual purpose: to keep the elements out and the conditioning of the home in. Sheathing is not only structural (giving the home strength) but also an integral part of waterproofing the house. As builders, we continually assess each project to ensure we maximize barriers that keep rain, wind, dust, pests, and pollution out of the indoor living environment.

We always build with the best when it comes to sheathing. There are several options we know, love, and trust—products that provide a safe and effective barrier and are very straightforward when it comes to proper installation. As with everything in this book, we will not recommend a material unless we are confident in its long-term performance and durability.

VAPOR BARRIERS

Before we go into the types of sheathing and their pros and cons, it's important to briefly discuss vapor barriers. A vapor barrier works like a shield, but not all vapor barriers are designed to block every type of vapor. For this chapter, we are referring to moisture vapor barriers.

There are several schools of thought regarding a moisture vapor barrier's place in the home and when to deploy them during construction. Whether you need to use a vapor barrier (and what kind of barrier to use) depends largely on the region in which you live. The US Office of Energy Efficiency and Renewable Energy lists eight different climate zones in the continental United States. If you live in zone four or higher, for example, class one and two barriers are recommended. Your builder will know more about this and your local energy code requirements.

One of a builder's primary jobs is to keep water out of the house—but, if water does get in, the house also needs to let water out. Water can take the form of a solid, liquid, or gas. Many in the building industry focus only on the obvious culprits by keeping out ice and water, the solid and liquid states. But water in a gaseous state (vapor) poses an equal but often more insidious threat. After all, we can see liquid H_2O, but water vapor is invisible.

Humidity can often build up in the wrong places when water enters a home as vapor. This moisture may become trapped inside the home and/or eventually condensate on a home's interior surface in liquid form. Condensation may even form in areas not visible to the homeowner, such as inside ductwork or behind the walls, which means the issue can go undiscovered for years.

If more than one vapor barrier is used on the exterior walls of a home, humidity will likely become trapped between them. This perpetually dark (and now damp) area is a perfect place for mold to proliferate. An eye-opening study conducted by the Harvard School of Public Health suggests that three out of four buildings or homes have either a current water problem or water damage from previous issues.[2]

WHEN VAPOR BARRIERS GO WRONG

Building Science Digest states that incorrect use of vapor barriers is leading to an increase in moisture-related problems. Although originally designed to prevent assemblies from getting wet, vapor barriers often prevent them from drying out instead, especially when installed on the interior. This can create problems in any air-conditioned or exterior space of the home. The culprits are primarily when a vapor barrier is installed twice, vented masonry is blocked, or brick is installed over building paper and vapor-permeable sheathing.[1]

condensation

Condensation is the opposite of evaporation. It occurs when water vapor turns back into a liquid state. This happens in one of two ways: the air either cools to its "dew point" or becomes so vapor-saturated that it can't contain any more water.

Air always holds some water vapor; its presence is called humidity. This moisture can enter a home in several ways. Beyond leaks from rainwater or burst pipes, the most common sources are the ground, cooking and washing, and human respiration and perspiration. That's right—you create moisture in your home just by being there!

One common place to see condensation is on your home's windows. These flat, clear surfaces don't necessarily cause condensation, but they do provide a good surface for it to collect. If the inner parts or panes of your windows show excessive "sweating," moisture is probably also gathering on your ceilings and walls.

Beyond properly installing vapor barriers, insulation, and ventilation during construction, you can take these steps to combat moisture in the home:

· Ensure clothes dryers and gas appliances are properly vented and the vents are unobstructed.

· Use exhaust fans in bathrooms and kitchens.

· If possible, open a window when cooking or bathing.

· Do not store firewood indoors; freshly cut wood contains water.

· Consider a whole-home dehumidifier.

In fact, researchers have identified that avoiding dampness and moisture-related indoor exposures is of primary concern for preventing asthma and other respiratory conditions among sensitized and unsensitized individuals alike. Before becoming sensitized, an individual can be exposed to elevated levels of the allergen without experiencing an allergic response, but once a person becomes sensitized, even low levels of the allergen may initiate a reaction. The onset of these types of allergic responses can be extremely difficult to reverse once they start and may end up lasting a lifetime for some people.

SHEATHING

The sheathing product we incorporate into nearly all of our construction projects is a two-in-one weatherproofing membrane. Although it is a structural and waterproofing membrane combined, the beauty of this type of sheathing is it allows any rogue vapor in the exterior walls to exit via tiny pores. For this reason, it is not technically considered a vapor barrier, which is good because we want our vapor barrier deeper inside the walls and closer to the living space. Furthermore, this two-in-one product offers an integrated taping system that ensures all seams, joints, and nail holes are covered and sealed.

The disadvantage of this two-in-one product is that the lumber backing is typically oriented strand board (OSB). Like particleboard, OSB is a type of engineered wood created by compressing layers of wood "strands" or flakes using adhesives and other chemicals, which can off-gas. To ensure no fumes off-gas from the sheathing, we have developed a protocol for vapor barrier placement followed by a process to prevent OSB, plywood, and lumber from off-gassing into the home.

The second sheathing style option is a two-step process using house wrap after the home is sheathed with lumber products. While high-quality house wrap serves a useful purpose—mainly keeping a home dry, comfortable, and protected—these wraps lack rigidity and tend to tear easily. Be careful during installation and flashing when using house wrap, as breaches in a home's waterproofing layer can lead to costly and detrimental mold and pest issues later. If you decide to go with house wrap, we recommend using untreated exterior-grade plywood.

Both these sheathing systems will serve as a waterproof barrier to protect the inside of the home from outside elements. However, what makes these technologies so fascinating is how they both keep water out while still allowing moisture vapor to pass through to the outside. So, as we mentioned before, although these systems are waterproof, they are intentionally designed not to serve as vapor barriers.

GEEK BOX

engineered woods

Composite wood products are made by gluing wood waste fibers together using chemicals like resin, then adding heat and pressure. Many builders use these materials because they are inexpensive and versatile. However, the industrial processes and ingredients used in their construction can often off-gas harmful VOCs. As a result, we do not use traditional composite woods inside the airtight envelope of our homes. Here are examples of the types of engineered woods used in many homes today.

Plywood: This engineered wood comes in two grades—exterior and interior. Exterior-grade plywood is used in subfloors and roof decking and sheathing, while interior-grade plywood is often used in furniture. Both use formaldehyde as a binder, but exterior typically uses lower amounts of phenol formaldehyde (PF). Ironically, interior-grade plywood often uses more urea formaldehyde (UF), which emits more formaldehyde over a longer period.

Oriented strand board (OSB): OSB is used in some sheathing for roofs, exteriors, and floors. For adhesives, it typically uses PF-based glues along with methylene diphenyl diisocyanate (MDI) and other harmful chemicals that off-gas compounds classified as potential human carcinogens.

Particleboard, aka low-density fiberboard (LDF): This type of compressed wood usually uses MDI and UF—the type that off-gasses more and for longer. It is often used in inexpensive furniture and cabinetry. We avoid all particleboard.

Medium-density fiberboard (MDF): You often find MDF inside cabinets, furniture, baseboards, and solid- and hollow-core doors, and like interior-grade plywood, it is usually made using UF. Certain brands offer formaldehyde-free adhesives, but then the question is what else was used as a binder and where the wood chips were sourced. For these reasons, we avoid MDF.

High-density fiberboard (HDF): Sometimes called hardboard or fiberboard, HDF can be made using paraffin wax, ammonia, and various types of formaldehyde. Doors, pegboard, and substrates for many types of floorings such as laminates are made of HDF. We avoid it in our homes.

FLASHING THE HOME

One of the most crucial steps for effectively sealing and waterproofing a home is to properly flash the house wrap or two-in-one sheathing. Flashing describes any thin layer of waterproof material that keeps water out. All joints, seams, and penetrated screw and nail holes should be covered with the proper waterproofing tape. Sheathing butts up against doors and windows as well, so flashing all windows, doors, and exhaust vents is imperative, as these areas are often weak links. We over-flash all our homes with no exceptions.

Sadly, we've seen many remodels with improper flashing (or even no flashing at all), and moisture penetration is often a direct consequence. For example, we did a remodel in Austin for a couple who purchased a home with all sorts of moisture problems which resulted in a terrible black mold infestation in their HVAC system. When our team discovered that the entire master bedroom exterior wall had no proper sheathing behind the siding, it baffled us. How does a home builder whose work must pass inspection get away with leaving off such a crucial piece of the wall?

SHEATHING AGAINST EMFS

Another note on sheathing considerations for a healthy home is neighbor proximity and unwanted EMFs entering the house. Close neighbors can easily and unintentionally "share" their Wi-Fi, Bluetooth, and other unwanted frequencies from electrical equipment and meters.

Our build team is constantly seeking solutions for our sensitive clients. Some have proposed using materials that will block electromagnetic frequencies from entering the home. While this sounds good in theory, it's not always practical for every homeowner.

A prime example of this happened years ago, when we built a home using aluminum foil–backed plywood on the sheathing and roof deck. While this did help keep frequencies out, there was an unwanted side effect: any phones, computers, car keys, or other emitting devices were magnified within the home. This may be a problem for some homeowners, so it's important to consider the impact of building materials from every angle. Thankfully, this homeowner was happy, as she was careful to keep EMF-emitting devices outside her home.

ENTRANCES AND EXITS

What's a healthy home without natural light, ample ventilation, and a way to get in and out? Not healthy or a home, that's what. When it comes to the exterior of a home, doors and windows can be found along all the walls and even on roof decks.

Exterior doors and windows are as aesthetic as they are functional, but they are still "holes" in your home's skin. It's just as important to maintain the structural and operational integrity of your exterior barrier as the quality of the indoor living environment. That said, a Google search for window and exterior door manufacturers can be overwhelming, with literally thousands of brands available. So, let's briefly discuss these holes in our homes to make sure they're done right.

First, find a reputable door and window manufacturer in your region. Equally important is finding an installation crew that is familiar with installing your specific door and window package. If you're thinking about saving a buck on more economical options that lack high-quality features, we highly recommend against it. You get what you pay for with virtually all construction materials, and subpar windows and doors will lead to expensive leaks, costly energy bills, and poor indoor air quality within a home for years to come. Keep in mind that some brands are only available in certain regions, mainly due to glass and frames being expensive to ship.

THE PERFECT WINDOW

There are several important factors to remember about windows. First, a true thermal break is essential for ensuring windows can efficiently keep cool air in and hot air out or vice versa. The thermal break is a pocket of air insulation within the window frame—a continuous barrier between the inner and outer frames that prevents thermal energy loss (aka thermal bridging, which occurs when heat or cold can easily transfer across a window frame without an air gap between the exterior and interior sides).

Of equal importance to thermal efficiency are the windowpanes. Look for double (or triple) pane glass filled with gas (usually argon). It's amazing how well this pocket helps insulate a home. Low-emissivity or "low-E" coatings are another great addition to aid in reflecting the sun's heat. Just be sure that any reflective treatments are applied to the enclosed inner or outside panes to avoid off-gassing into the home. Wind and water ratings are also good to review when comparing manufacturers.

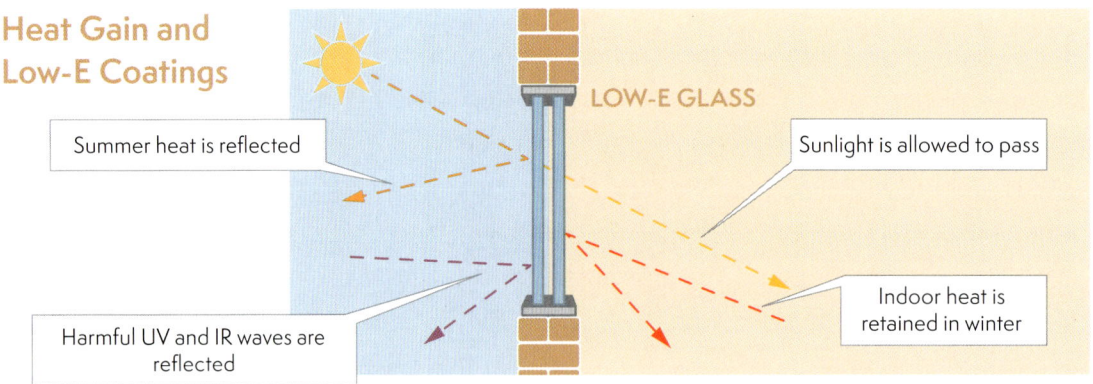

Heat Gain and Low-E Coatings

LOW-E GLASS

Summer heat is reflected

Sunlight is allowed to pass

Harmful UV and IR waves are reflected

Indoor heat is retained in winter

thermal bridging

A true thermal break should separate the interior and exterior frames and is as important for retaining heat in the wintertime as for keeping cool air indoors during the summer. Without this break, thermal bridging may become problematic, especially during the cold winter months.

All metals, including aluminum, transfer heat efficiently, which will lead to indoor heat escaping and a cold interior metal frame. Warmer air holds more moisture. Thus, moisture in the warm indoor air can lead to condensation on the cool interior metal surface. Condensation on window surfaces leads to unwanted mold and window warping, neither of which are conducive to a healthy home.

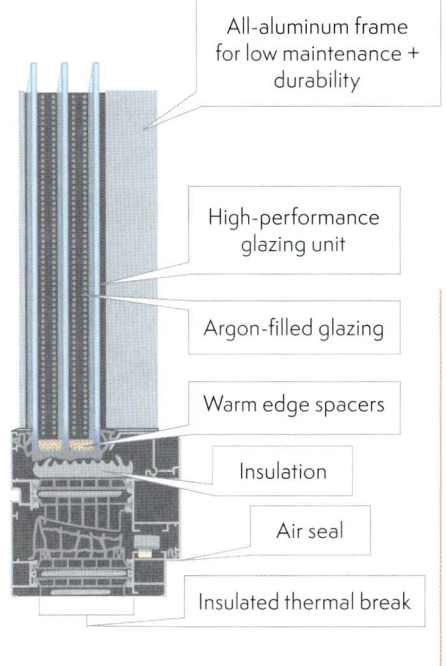

All-aluminum frame for low maintenance + durability

High-performance glazing unit

Argon-filled glazing

Warm edge spacers

Insulation

Air seal

Insulated thermal break

There are dozens of styles of windows and hundreds of ways to customize them. However, there are only a handful of window frame materials. Aluminum, iron, and steel are the best options in terms of durability, longevity, and nontoxic attributes. These are usually the most expensive options, with aluminum-clad windows being the least pricey of the three. However, if there's one place to put your money, investing in high-quality windows manufactured by a reputable company that carries a solid warranty is a good place to start. A true thermal break and high-quality panes are especially important for aluminum-clad windows to avoid thermal bridging.

Some reputable window manufacturers also produce window frames with metal exteriors and hardwood interiors, which are a good option. We order our metal-clad wood windows unfinished and unprimed on the interior so that we can apply our own nontoxic interior paints, stains, and topcoats.

High-quality residential windows should function well for the life of the window, as they are intended to keep water, pollution, humidity, and extreme temperatures out while allowing the conditioned air to remain inside.

Most residential homes constructed in the past couple of decades, however, were built with vinyl windows, which have become an unfortunate industry standard. Vinyl windows are cheaper to manufacture but can off-gas phthalates and PVC fumes into the home, especially during hotter months. Their exterior color also tends to fade after several years of sun exposure, which means the window components may need to be painted or replaced later. For these reasons, we do not build healthy homes with vinyl windows.

Ask your window manufacturer or supplier what the internal components of their windows are made of, since some higher-end windows use concealed vinyl components on the interior.

Also ask whether the manufacturer applies pesticides to its bug screens and whether those screens are installed on the inside or outside of the specific make and model you are considering. If they are on the inside, inquire about metal versus PVC screens without insecticide.

As an unfortunate side note, most mass-produced windows have pesticide treatments applied to the internal components during the manufacturing process. To date, we have not found a reputable window company that does not do so. Luckily, it is a small amount in an enclosed space and has yet to bother even our most sensitive homeowners.

GEEK BOX | PVC, phthalates, and your health

Polyvinyl chloride (PVC) is a nasty chemical that can off-gas for decades or even indefinitely as it degrades over time.

Exposing indoor PVC screens or vinyl windows to heat and UV light will dramatically increase the release of phthalates, harmful VOCs known to cause respiratory issues and act as endocrine disrupters once inhaled. Endocrine disrupters are chemicals that mimic or interfere with human hormones and can cause problems with our developmental, reproductive, immune, and other systems.

We've had some success with placing PVC screens in the sun for several months during construction to speed up the initial off-gassing potential, but we still recommend avoiding their use indoors whenever possible. Some higher-end window manufacturers offer roll-up screens that are easily removed and stored when not needed. Remember, there's always the option to remove most of the interior screens and keep a few on select windows.

SOLID WOOD, COMPOSITE, OR FIBERGLASS?

Other options do exist in the world of windows. While solid wood frames can be beautiful, the exterior frame will shrink and swell over time due to swings in moisture and heat. This can lead to paint chipping, wood rot, and issues with window functionality.

When fiberglass window frames first came out, they were available only in white and had issues with becoming faded and chalky over time. Recently, more colors have been introduced. However, their performance and composition remain uncertain at the time of this publishing.

A few manufacturers produce less expensive composite-based windows that we have found to be healthy options. While the materials in these windows are not completely formaldehyde-free, there is very little off-gassing. It's worth contacting the manufacturer to get details on the specifics of the composite.

As discussed earlier, ample roof overhang is important, and the same overhead coverage concept applies to windows. Pay particular attention to windows that are not covered by an overhang. For example, some houses have a first-story window located directly below a second-story window with nothing but flat wall between them. An awning or covering over the first-story window is a good idea to keep wind and driving rain from potentially breaking through after years of weathering. It can also help with solar gain.

When it comes to horizontal skylights, the answer is simple—no. We do not install them, as the risk for water leaks is too high. There are plenty of ways to design a home filled with natural light without placing horizontal skylight windows on the roof plane.

Seasonal Solar Gain and Window Overhangs

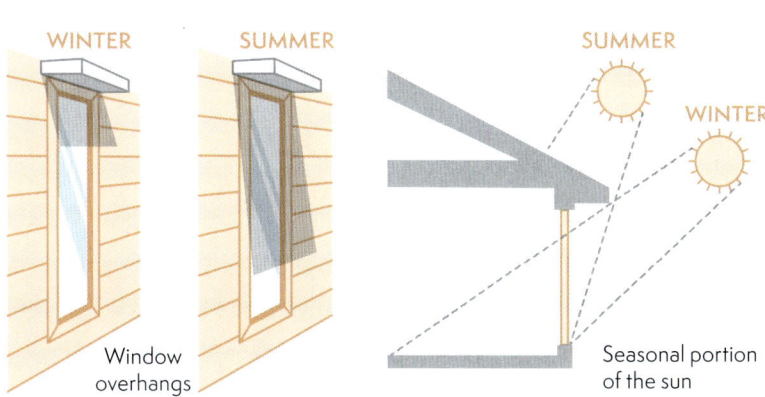

WINTER   SUMMER   SUMMER   WINTER

Window overhangs

Seasonal portion of the sun

DON'T FORGET DOORS

Much of this information applies to doors as well, but weather-stripping options are critical to consider. Aftermarket weather stripping for some exterior door models is a good idea. Be sure your weather stripping covers the base of the door.

Though not always feasible (for example, equipment may be in the way), it's best for exterior doors to open inward into the home because the door frame adds an additional layer of moisture protection from driving rains. We also recommend triple door lock mechanisms to boost security, rigidity, and tightness, especially on taller doors.

We touched on flashing earlier, but we must reiterate the importance of proper and adequate flashing around doors. This must be done during installation and requires the installer to use insulation in the surrounding cavities.

While builders often offer exterior window and door packages to ensure a consistent look and feel throughout the home, there are a few important considerations when choosing among brands and models. Among these are the factory warranty, the availability of installers in your area who are familiar with the manufacturer's products, and the R value (insulation capacity), because different types of windows and doors are sometimes manufactured for specific climates.

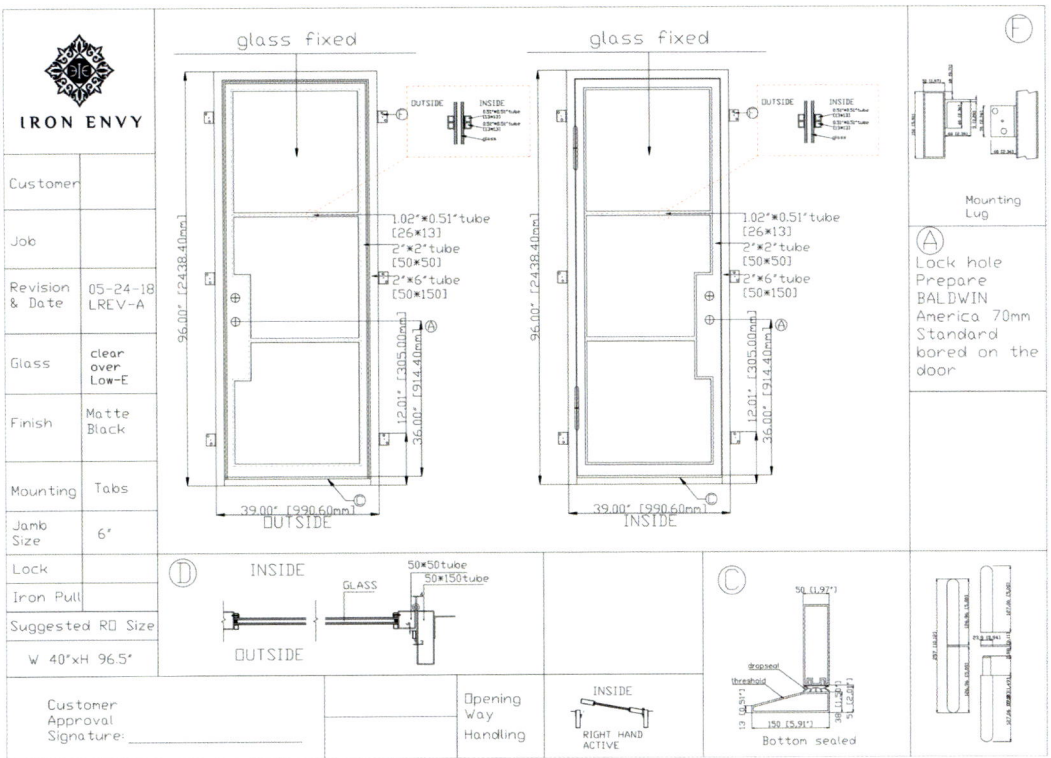

When it comes to finding high-quality exterior doors, the considerations are similar to those listed earlier for windows, including sourcing doors that offer a true thermal break. Check out the illustration above of a twelve-gauge steel frame and glass front door from one of our recent builds. Notice the red boxes calling out the sealed double-pane windows, which provide thermal and sound insulation. Box C illustrates the weathertight seal against air and moisture and the raised threshold, which acts as a water barrier. The low-E coating helps block solar rays and heat, as does the foam insulation that is completely enclosed inside the steel frame. A well-built front door is a solid investment!

THE FINAL LAYER

While we're discussing a home's exterior, consider this: if sheathing is a house's skin, masonry or siding is its clothing. After all, our wardrobe is meant to keep us comfortable, clean, dry, and of course stylish!

There are many exterior material options for houses. One of the most popular categories is masonry, which includes brick, stone, and stucco. Of those finishes, stucco is probably the most common worldwide, especially in more temperate climates.

We recommend using "real stucco" over cheaper alternative materials such as synthetic stucco or an exterior insulation and finish system (EIFS). EIFS was incredibly popular in the 1990s but was later found to be a recipe for wet and moldy disasters. In fact, most insurance companies will no longer cover a home that has an EIFS exterior. A dead giveaway of EIFS is paint peeling away from the exterior to expose a Styrofoam core.

We only use real stucco, a simple mixture that includes Portland cement, sand, lime, and water. The concoction is mixed into a thick goop and then applied over a metal lath or mesh base on the exterior of a home. Modified real stucco products with integrated paint colors are also available and offer additional waterproofing properties.

Stucco has been around for centuries as both a construction material and an artistic medium. When it comes to modern-day stucco applications, however, there are important installation guidelines to remember, such as spacing the stucco properly from the ground, using specific backing materials, sealing the exterior, and employing drainage best practices. These considerations will help ensure your home remains safe and healthy for many years.

Brick and stone are also very popular materials for home exteriors. Both have been around since the dawn of civilization, and for good reason. They stand the test of time and are readily available around the world.

When sourcing stone, denser options are always better than porous ones if they're available in your region. While denser stone is heavier, it also provides better moisture protection and is less inviting for mold and pests. Remember, stone that is mined locally is usually significantly less expensive than rock that has to be transported over long distances.

This applies to brick as well, which tends to be more porous than stone. Annual applications of a nontoxic spray-on waterproof coating are important to maintain water resistance on exterior brick walls. This coating can also be applied to exterior mortar, the paste that binds building materials together. Mortar between brick and stone is one of the most vulnerable parts of a home's exterior. Be sure to repair mortar cracks as they appear over time.

OTHER OPTIONS

Fiber cement planks and solid wood are also great choices for exterior siding and soffits. Fiber cement board and composite siding is readily available and comes prefinished or paintable. It's a great product as long as it is installed correctly and the exterior is maintained. There are specific requirements for each product for tongue-and-groove as well as shiplap installations. As with most exterior wood products, periodic maintenance is required to ensure wear and tear doesn't lead to leaks or insect infestations. Both fiber cement and composite should last a long time with proper drainage and regular touch-ups to the caulk and paint.

Solid wood is a step up from cement board. It's gorgeous but a bit pricier. Exterior wood cladding and ceiling features that use a tongue-and-groove design can offer one-of-a-kind character as well as long-term durability. Just make sure to keep up with periodic staining and sealing to keep it looking like new. Some brands even come prefinished and require very little maintenance.

Metallic sheeting is yet another siding option that has become more readily available in recent years. We do not recommend large areas of metal siding for two reasons. First, petroleum-based lubricants are generally applied to the expansion joints. Second, there is an inherent risk of amplifying EMFs within the home's envelope. Small decorative areas, however, should not pose any issues.

Large-format ceramic and porcelain sheets are also gaining in popularity, proving to be bold and interesting exterior options. That said, they are expensive and generally used on commercial buildings. Although we have yet to build a home's entire exterior using porcelain or ceramic tile, we recommend making sure that the tile you choose doesn't have an interior metal plate. In short, large-format porcelain cladding is definitely on our list of faves.

Before making any decisions on siding, we advise you to do your due diligence and research the best options for exterior construction in your region of the world. Doing so will ensure your home's exterior is both beautiful and durable for many years—with proper maintenance and tender loving care, of course.

key points

▶ Like roofing, exterior sheathing keeps elements out and conditioning in.

▶ Vapor barriers are important tools for keeping water out of a house.

▶ Avoid stacking vapor barriers on walls. One is the magic number.

▶ You can use a house wrap or a two-in-one weatherproofing membrane.

▶ Proper flashing with waterproof material is one of the most crucial steps for effectively sealing up a home to be airtight and watertight.

▶ When looking at windows, remember to find ones with a true thermal break and high-quality double or triple panes filled with argon gas.

▶ There are plenty of options for window materials; avoid vinyl panes and components, PVC, and other materials that can emit chemical fumes.

▶ For doors, we recommend proper weather stripping, a solid frame, and triple lock mechanisms for added security and tightness.

▶ For the exterior, go with real stucco, stone, brick, composite, or wood—just make sure to perform regular maintenance.

Up next: what are the keys to maintaining health within your indoor living environment? Read about healthy indoor construction in the next chapter, *Interior*.

interior

We know it's important to eat a healthy diet, take vitamins, and consume the right number of calories to fuel exercise and boost cardiovascular health.

Many of us are willing to pay a premium for organic whole foods and monthly gym memberships. Some of us pay thousands of dollars for workout equipment to use at home. We even spend top dollar on premium-priced organic clothing and sheets—all in the name of healthier living.

This is all great, beneficial, and essential to living a better life. However, it's truly surprising how few people stop to think about the bigger picture: their indoor living environment.

WHERE WE LIVE + WORK + PLAY

Why do so few people consider the quality of their indoor living and working environments? These are the places where we eat, sleep, and breathe. We spend 90 percent of our lives inside our homes, offices, and schools. It's kind of a big deal.

As builders, we start by considering how changes in the outside climate affect the indoor environment, accounting for extreme weather, temperature fluctuations, air quality issues, and more. Thermodynamics play an important role in how we design exteriors to protect the interior living space. In simple terms, weather changes constantly, whereas indoor environments should remain stable and comfortable.

Obviously, different climates and seasons change which side of your wall is considered the hot or cold side. If you live in Canada and it's wintertime, the inner part of the house is the warm side. On a hot, dry, sunny summer afternoon in central Texas, however, you could probably fry an egg on the exterior walls. The environment outside the home and the conditioned spaces within the home are both dynamic—thus, thermodynamics.

An important concept for building boils down to what we call thermal breaks. These are important because water vapor tends to migrate away from warm areas and condensate on colder surfaces. It's like starting your car on a cold winter day. You get in, crank up the heat, and watch as the cold windows

inevitably fog up. However, if you turn on the defroster, it heats the windows and stops condensation from forming.

We want to avoid extreme thermal differences between a home's exterior walls and roof decking and its interior core. We achieve this by creating a thermal break around the entire exterior shell of the house. How? Through adequate insulation.

INSULATION AND THE THERMAL BREAK

Insulation is a hot topic around our construction office. We approach this part of the building process differently than other builders do.

For starters, we generally go beyond minimum local building codes and regional US energy requirements to meet our healthy home criteria. Sometimes going above and beyond on insulation is a good thing. No matter where you live, outside elements can and will become extreme at different times of the year. Often people think of insulation as keeping heat in and cold out, but in many places, it works just as hard during the summer to achieve the opposite effect.

We live in central Texas, where late-afternoon air temperatures in the summer can exceed 110 degrees Fahrenheit. As the sun sets, the elevated ambient air temperature combines with the sun's intense rays to heat the exterior surfaces of a house to more than 150 degrees.

Proper insulation creates the thermal break, and the best insulation is actually air! For decades, homes have been insulated with fiberglass batts, which are dense blankets that trap air in millions of fibers. These batts are relatively cheap to mass-manufacture and made of materials that neither conduct nor radiate heat.

In the past couple of decades, batts made from other materials have become available, such as mineral wool. While neither mineral wool nor fiberglass serves as a suitable food source for mold (definitely a plus), there is a common misconception that mineral wool is a healthier alternative.

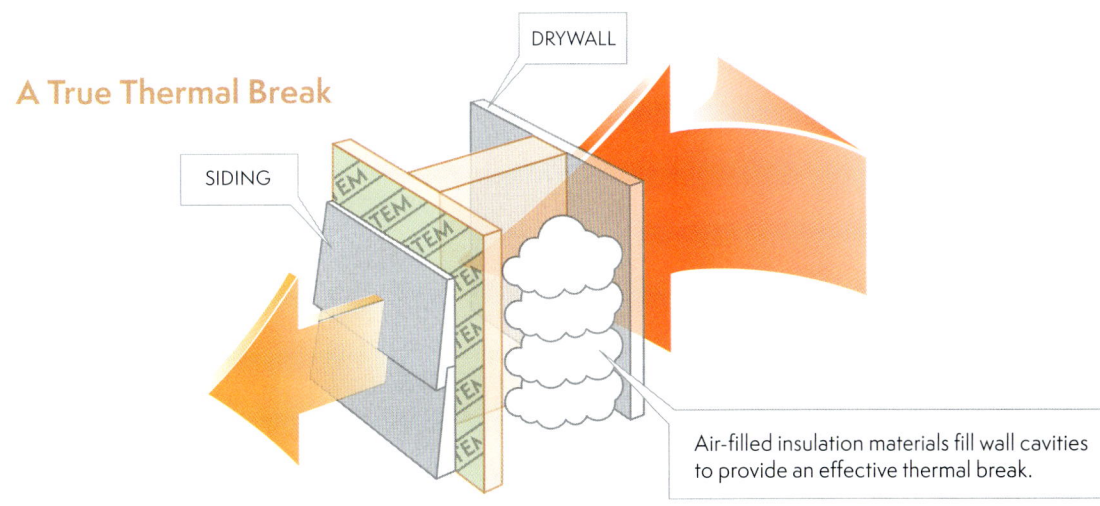

A True Thermal Break

DRYWALL

SIDING

Air-filled insulation materials fill wall cavities to provide an effective thermal break.

The issue with most of these batt insulation products is the materials used during manufacturing and packaging. The binders in the batting may contain formaldehyde and solvents, which will continually off-gas into the home. Furthermore, the paper-backed insulation sheets contain adhesives that are derived from bitumen (nasty tar binders).

If you've ever been near a freshly asphalted road on a hot day, you know the kinds of fumes petroleum-based tar products can create. Lining ceilings and walls with these bitumen-laced materials can be detrimental to the home's indoor air quality. Sadly, almost all homes with batt insulation have petroleum-based bitumen in the walls.

As if the binders in batting weren't concerning enough, there's also the issue of paper or kraft backing. We were excited when select manufacturers introduced new formaldehyde-free fiberglass and mineral wool products several years back, but it is our job as healthy home builders to do our due diligence. Upon further investigation, we found that the adhesives used to bind the batting to the paper backing were still asphalt-based across all formaldehyde-free brands currently on the market.

Effects of Chronic Exposure to VOCs

CNS disorders (e.g., styrene)

Neurological effects (e.g., toluene)

Pain sensitivity decrease (e.g., TMB)

Eye irritation (e.g., formaldehyde)

Skin irritation (e.g., pentane, mycotoxins)

Hearing problems (e.g., xylene)

Irregular heartbeat (e.g., benzene)

Lymphocyte count reduction (e.g., benzene)

Irritation of respiratory tract (e.g., formaldehyde)

Liver and kidney toxicity (e.g., polybrominated diphenyl ethers (PBDE)/styrene)

Kidney weight increase (e.g., toluene/cumene)

Fetal weight disorder (e.g., cyclohexane)

Myeloid leukemia (AML) (e.g., benzene)

Developmental toxicity (e.g., PBDE)

Abdominal problems (e.g., methylpentane)

Peripheral neuropathy (e.g., hexane, mycotoxins, heavy metals)

If you decide to use blanketed batts with no backing, be sure to wear protective gear and to vacuum the home well after installation to pick up stray fiberglass filaments.

Over the past two decades, the construction industry has introduced various blown-in insulation products as alternatives to rolled batts. All sorts of materials can be blown into a home's interior cavities, including expanding foams and other substances.

For air-blown materials, the process typically involves stapling a plastic sheet to the interior studs, which creates a cavity that can be filled using a hose inserted at the top. Shredded recycled denim jeans are among the more creative nontoxic materials used.

Years ago, we tried a blown-in product that utilized a nontoxic magnesium oxide–based foam. This mineral-based formula was inert, had a good R value, and foamed up quite well, which seemed especially promising. Unfortunately, the result lacked structural durability and consistency upon curing. We hope manufacturers of innovative products such as mag-ox insulation continue to refine product performance in the coming years so that someday soon they may be superior in durability, performance, and composition.

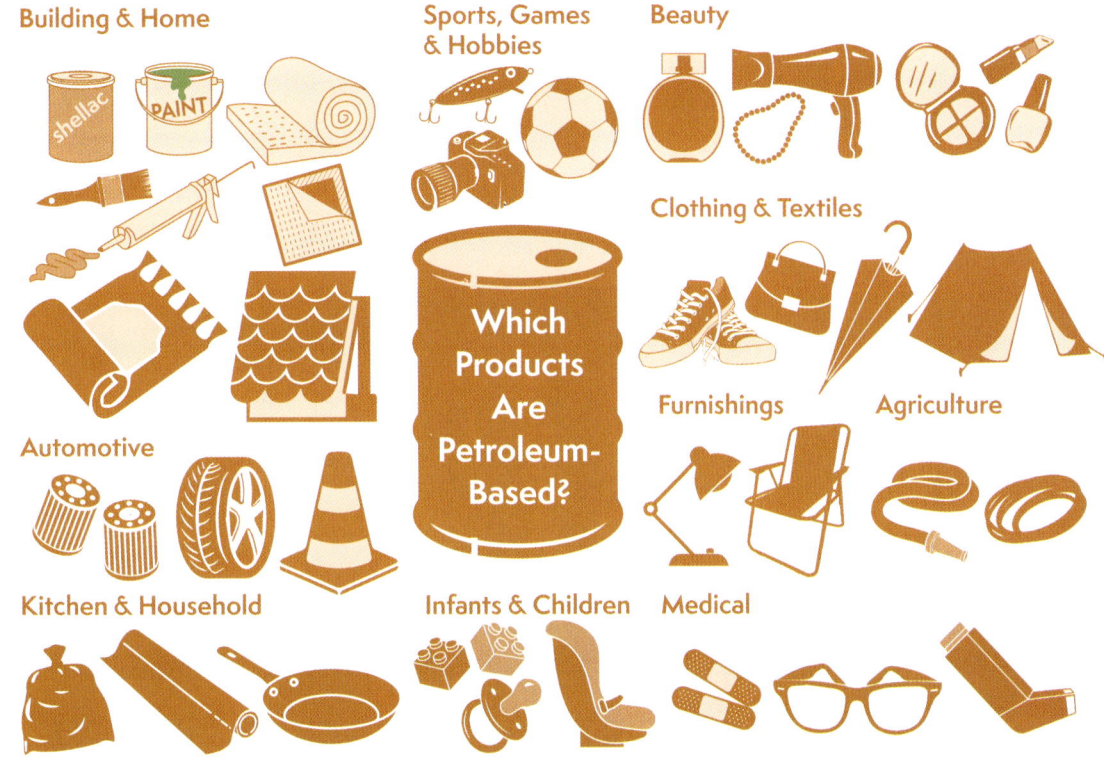

what is an R value?

R values measure an insulation's ability to resist the conductive flow of heat—that is, how well it keeps heat from leaving or entering a home. This is usually measured per inches of thickness, although R values vary based on the insulation type. Blown-in insulations tend to have lower R values, while batts and foams have higher values. Higher values usually mean better climate control, but they also tend to mean "more expensive."

The manufacturer generally lists the R value on each of its products. A professional contractor will utilize the appropriate software to run calculations that will ensure your home conforms to regional energy code requirements. Check the energy code map on the Healthier Homes website for up-to-date US federal requirements by region and to figure out what R values you need for your home's materials. Remember to check local and city building codes, too, as these are constantly evolving and may supersede other requirements.

CONCERNS ABOUT ICF

Another alternative is polystyrene-based insulation (think Styrofoam). Not only does insulated concrete form (ICF) construction utilize expanded polystyrene—the use of which we do not condone in healthy home construction—but it also has its own share of problems from an occupant health perspective.

Expanded polystyrene (EPS) is available in an interior wall insulation application and standard insulated concrete form construction, and both will inevitably off-gas styrene, a well-known and toxic VOC. Humidity and VOC concerns also apply to alternative ICF products, which use a slurry of recycled earthen materials of varied and often unknown origins.

In addition to being more expensive, requiring more floor space, and being difficult to remodel, faulty ICF construction that uses polystyrene forms can allow groundwater and insects such as termites into the walls. These forms also tend to increase indoor humidity, especially while the sandwiched concrete is curing. All of this creates a potential breeding ground for mold and fails to address the need for a vapor barrier.

FOOD FOR MOLD

While products like recycled denim are natural and "green" alternatives, we have two caveats. First, the insulative value of denim may be inconsistent, which is why you should always ask about the R value of your insulation product. Second, cotton can serve as a food source for mold growth in the right conditions.

While it's best to avoid products that serve as food for mold, the reality is that homes do utilize many carbon-based materials such as lumber, cellulose derived from wood (paper), and even cement, all of which can grow mold in the right conditions.

An integral part of our healthy home building strategy is controlling indoor conditions. This includes keeping moisture carefully contained and humidity in check. Stay tuned for more details on indoor air quality in the next chapter.

ICF Wall with Stucco Finish

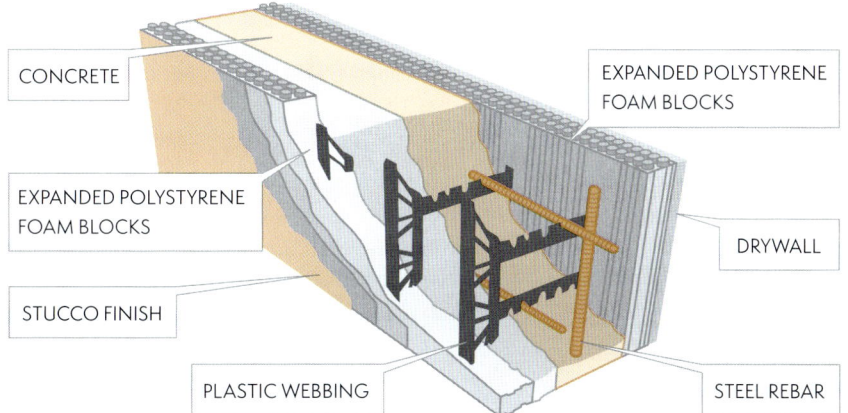

CONCRETE

EXPANDED POLYSTYRENE FOAM BLOCKS

EXPANDED POLYSTYRENE FOAM BLOCKS

DRYWALL

STUCCO FINISH

PLASTIC WEBBING

STEEL REBAR

OPEN CELL SPRAY FOAM

For the past decade, the industry standard has been spray-foam insulation due to its ease of application, high R value, and ability to cover large surface areas adequately. The foam expands upon contact to penetrate the nooks and crannies within the framing, sheathing, and roof decking. Just like batting, denim, and polystyrene, spray foam creates an air barrier through a matrix of tiny bubbles within the foam.

However, not all spray foams are created equal. We are very specific about the spray foam we install, as most products are highly toxic and will off-gas for years. Our means and methods differ from other builders' when it comes to spray-foam applications, and the materials are just as important as the application process.

But first, for ease of comparison, let's establish the two types of expansive spray foam: open cell and closed cell.

Open cell spray foam is the more economical choice for the vast majority of residential builders. With enough layers, the R value meets minimum code requirements, but it is also a highly toxic concoction that cures in a soft, spongy fashion with its cells wide open. The result is perpetual off-gassing of countless VOCs and an alarming amount of formaldehyde buildup within a home. It breaks our hearts to think of the number of people with open cell foam behind their walls.

Years ago, we faced quite a challenge when changes to energy code standards meant that, collectively, we as builders could no longer meet regional energy code requirements using batting insulation in exterior walls. Additionally, many jurisdictions began requiring blower door tests. (Remember, these involve attaching a giant vacuum to an exterior door while monitoring air pressure on the interior to measure air leakage.) We had to determine how to build an airtight home that complied with stringent R value requirements without compromising indoor air quality.

Open cell spray foam was out of the question due to its polyurethane base and open cell structure, which allows petroleum-based chemicals and solvents to continually emit noxious fumes. But we found an answer.

CLOSED CELL SPRAY FOAM

Closed cell spray foam was our solution to keeping up with increased R value requirements while maintaining a healthy living environment inside the home. While closed cell foam is more expensive than open cell and is traditionally used in commercial building, we decided years ago to use it in our residential projects. Its performance is far superior.

This product has twice the insulative R value per inch, and its closed cell design means that once the foam has cured, the fumes from the isocyanate-based polymer resin are encapsulated. While some off-gassing will occur, the amount and duration are drastically less compared to open cell spray foam.

This spray-applied plastic foam has a low expansion rate, is known to resist heat transfer extremely well, and forms an air seal wherever it is installed, which reduces air infiltration and helps lower monthly energy bills. Even better, it's basically moisture-proof. It doesn't retain water or allow moisture through, meaning mold doesn't stand a chance against closed cell spray foam.

DRYWALL APPLICATIONS

Drywall, also called sheetrock, is one of the more controversial construction materials. Many of our homeowners are skeptical about using it. In fact, one homeowner asked us if we could build his entire home out of hard, nonporous surfaces, including the walls and ceilings.

Perhaps homeowners are right to be concerned. After all, the walls and ceilings are the largest surface areas within a home. Plus, in the early 2000s, drywall products imported from China got a lot of bad press when high levels of sulfur were found to be used in manufacturing. Homes built with these products had a persistent rotten egg smell, and a host of health problems were attributed to chronic exposure, including asthma, coughing, headaches, and eye and skin irritation.

Many clients are also concerned about drywall's supposed propensity for mold growth, but this occurs only in conditions of constant moisture from high humidity or a leak. Controlling the moisture in an indoor environment will eliminate the issue of mold on drywall.

That said, most commercial drywall contains a number of toxic additives, including formaldehyde, paraffin, biocides, and various other chemicals that release harmful VOCs. But perhaps the most concerning fact about most of these products is that commercial gypsum is primarily sourced as an industrial by-product from coal plants. Fly ash waste is scraped and "repurposed" into synthetic gypsum, which can even be radioactive.

We source only natural gypsum drywall, with zero formaldehyde, sulfur, or any other chemicals that will off-gas.

Wallboard made from alternative materials such as magnesium oxide may be nontoxic, but most of these products are currently manufactured in Asia, and tracing the materials' origins can be quite challenging. The installation process for magnesium board is similar to that for plaster, requiring a skilled installer. While all forms of drywall will eventually develop a few small cracks along the seams, perhaps the biggest pitfall with magnesium board products is that they don't allow for natural movement and deflection in the walls, which occur as homes shift or the outdoor temperature swings. This can lead to drastic cracks in interior walls.

These principles also apply to the tape-and-float process, which involves creating a solid joint between two drywall pieces so that plaster and paint don't crack later. We use a joint compound (often referred to as "mud") that contains no formaldehyde, harmful additives, or fungicides. Since the mudding process requires the use of water, the home must have adequate air flow to allow the mud to dry. Don't forget to use clean, potable water.

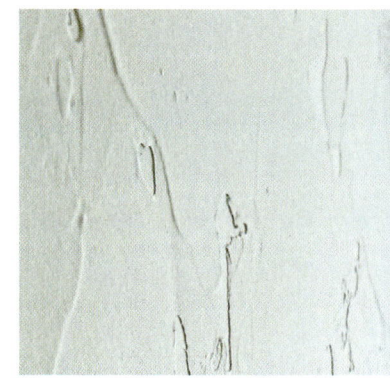

Some nontoxic plaster products for walls are made from clay. These are beautiful and relatively nontoxic, but be aware that the multi-step installation process is highly specialized and a bit costly. We recommend avoiding the use of clay products on walls in wet rooms such as bathrooms with showers.

When finishing the interior of a home, remember that the garage and attic spaces are just as important. Some builders skimp on insulating or drywalling the garage, mechanical rooms, and attic spaces, but we treat them just like interior living spaces. After all, the air circulates throughout the entire home.

For more information about paints and other interior finishing options, see Part 4.

key points

▶ The climate outside your home affects the indoor environment, which is why thermal breaks are so important.

▶ There are lots of choices when it comes to insulation, so pay attention to the material composition, R values, and local building energy code requirements.

▶ When it comes to drywall, do your research and go with natural products. Just ensure the home's humidity is well regulated.

▶ All spaces within a home's envelope are equally important when it comes to the use of healthy materials and code-compliant building practices.

In the next chapter, we'll discuss innovative approaches to ensuring that the air in your home stays fresh and clean.

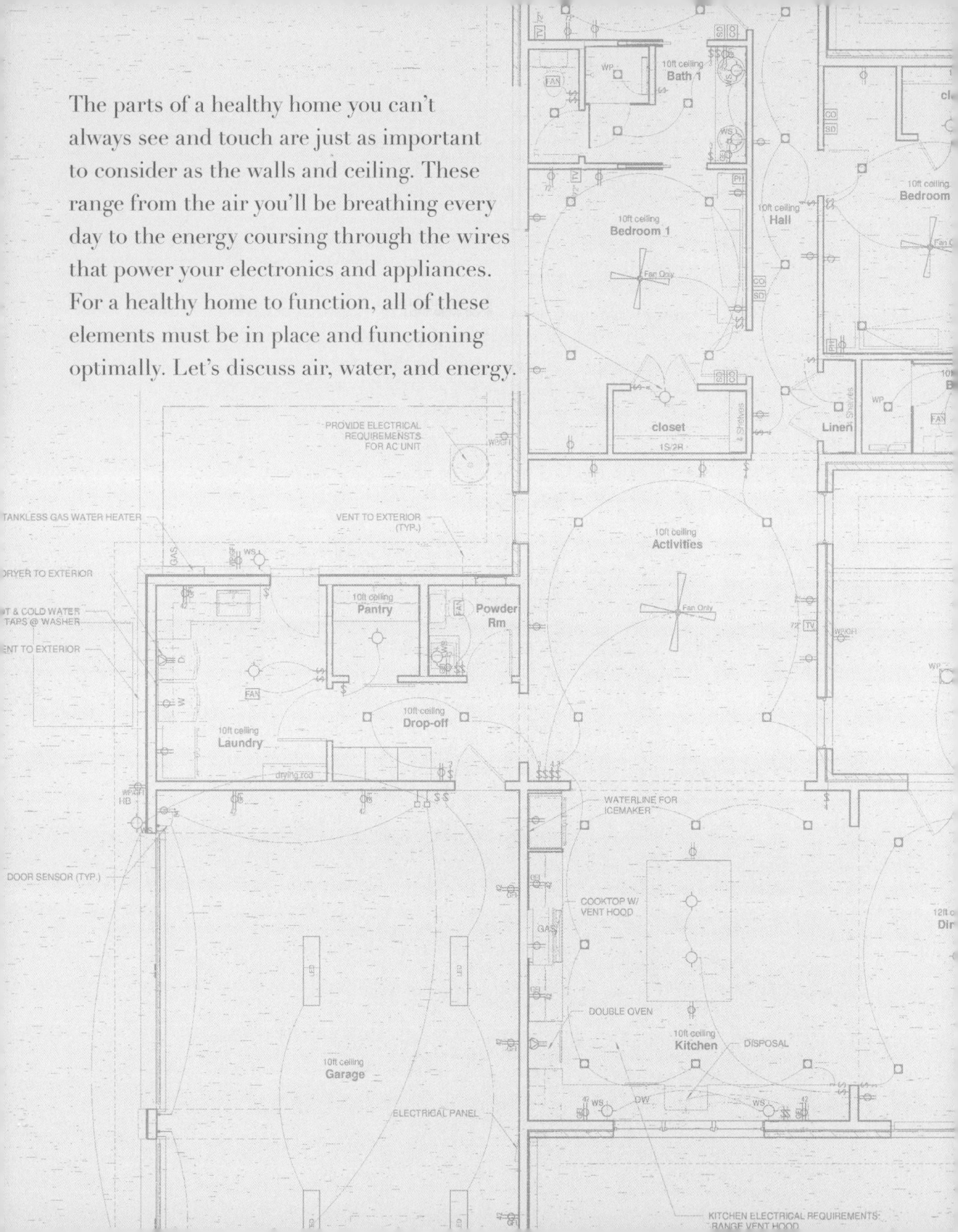

The parts of a healthy home you can't always see and touch are just as important to consider as the walls and ceiling. These range from the air you'll be breathing every day to the energy coursing through the wires that power your electronics and appliances. For a healthy home to function, all of these elements must be in place and functioning optimally. Let's discuss air, water, and energy.

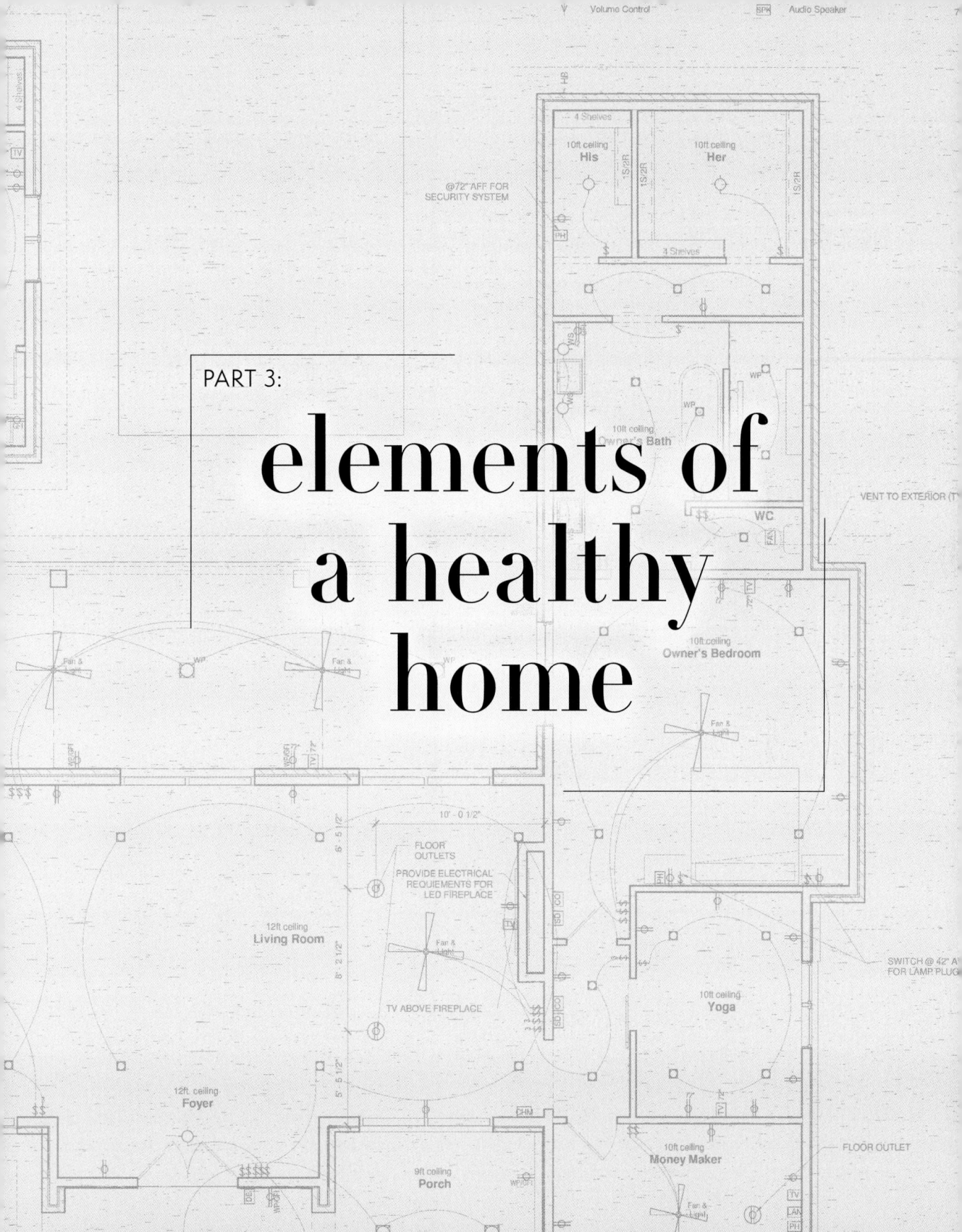

PART 3:

elements of a healthy home

air

By forcing so many people to stay home, the COVID-19 pandemic brought indoor air quality to the forefront of everyone's minds. As a result, many residential HVAC manufacturers have begun advertising home sanitization systems. Plus, many businesses are starting to see the value of keeping germ counts inside their buildings down, meaning commercial air purification systems are suddenly all the rage as well.

Despite the unfortunate circumstances, we're excited that indoor air quality is finally starting to get noticed. We have been installing whole-home electronic air purification systems for years, including ozone generators, UV lights, and HEPA filtration systems.

AIR PURIFICATION SYSTEMS

When most people think of air purification, they think of air filters. The gold standard is the high-efficiency particulate air (HEPA) filter, a type of pleated mechanical filter that removes 99 percent of dust, pollen, mold, and bacteria.

Spectrum-specific ultraviolet (UV) light can also purify air. Unlike HEPA filters, which simply trap particles, UV filtration uses ultraviolet light technology to damage viruses and other microorganisms in ways that keep them from reproducing. This can stop illness-inducing microbes, but it doesn't work on most other particulates, so the best UV light systems often use filters, too. They also do not replace traditional cleaning.

Properly installed ozone generators are completely safe. They produce a small and harmless amount of ozone, which is highly effective at eliminating odors and airborne germs, spores, and other contaminants. The generator takes oxygen (O_2) from the air and gives it an electrical charge to create ozone (O_3) molecules—the same process that occurs naturally in the Earth's atmosphere. The O_3 molecules attach to pollutants at the molecular level and destroy the cell wall, and then lose their extra "O" to revert to oxygen. Make sure your HVAC specialist knows how to install such a system correctly so that the generator doesn't keep running when the HVAC is idle. This would cause a buildup of ozone within the system, which can be irritating for some people.

Modern Air Circulation Systems

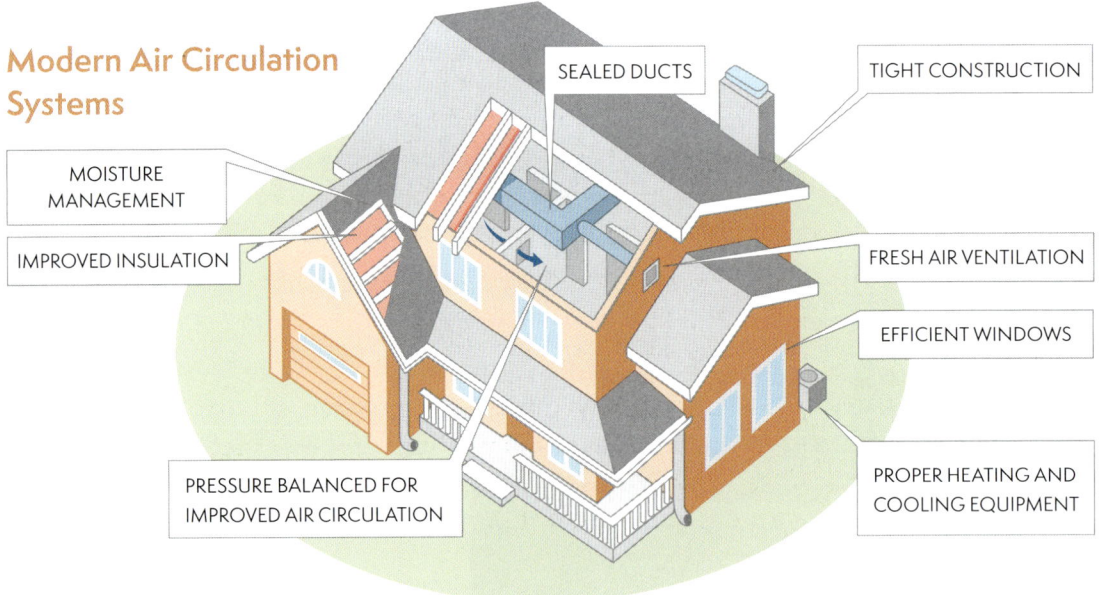

MOISTURE MANAGEMENT

IMPROVED INSULATION

SEALED DUCTS

TIGHT CONSTRUCTION

FRESH AIR VENTILATION

EFFICIENT WINDOWS

PRESSURE BALANCED FOR IMPROVED AIR CIRCULATION

PROPER HEATING AND COOLING EQUIPMENT

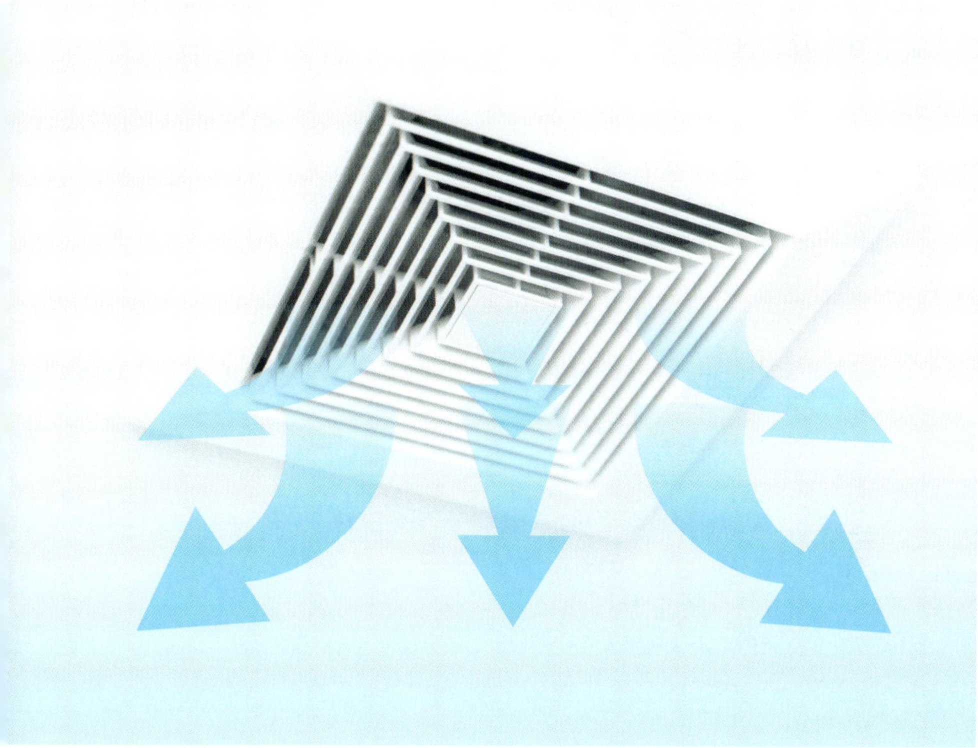

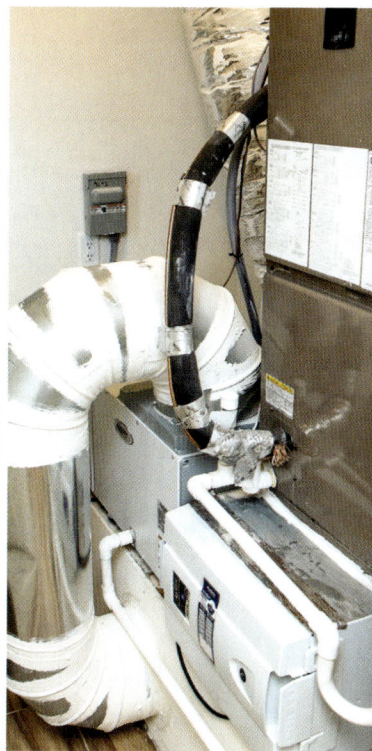

GETTING SOME FRESH AIR

As we've discussed, modern US building codes require contractors to adhere to specific energy standards. Builders often must make homes so airtight that only a few pinholes of air are allowed to escape during a positive air pressure test. This is why the air inside a home must be kept fresh, clean, and free of VOC buildup.

Fresh air ventilation should be a part of all HVAC installations. Keeping your fresh air intake at 50 percent is ideal. We recommend installing a dampered system (which uses a valve or plate in a duct that regulates airflow) on your HVAC to allow control over how much exterior air enters the home and when.

Be sure to place your fresh air intake and exhaust vents in appropriate locations. For example, intakes should be positioned away from exhaust vents and other pollution sources and be protected from rain and pests, while exhaust vents should be located where snow or debris will not obstruct them.

CIRCULATING AIR AND AVOIDING MOISTURE

HVAC system manufacturers have been hard-pressed to cut back on energy consumption, which has resulted in units that cycle a limited number of times per hour. While this is great for conserving energy, the lack of circulation can cause moisture vapor to build up inside a home. Make sure that your HVAC system allows you to override the number of cycles per hour in case your home starts to feel stuffy.

HVAC size

For HVAC units, bigger isn't necessarily better. Oversizing your HVAC can result in a variety of problems, including

- **Stressing your unit:** A huge HVAC system in a tiny house will cool the space very quickly, which stresses the unit as it turns on and off constantly. This can also decrease cooling efficiency, increase energy bills, and cause condensation.

- **Boosting humidity:** A/C coils create condensation, which goes into a drain line or drips out the side of a home. If condensation builds up too quickly, there's a risk of overflow into the drain pan. We install moisture-sensing shutoff switches for these pans as an extra precaution.

- **Reducing air quality:** As we discussed, every AC unit has intake and exhaust vents, but an oversized unit—which causes shorter cycles and humidity buildup—can lead to poor circulation, reduced air quality, and the formation of mold.

If you're experiencing excessive condensation on windows, mold growth, or a mildewy smell when the unit turns on, your problem might be an oversized HVAC unit or dirty coils.

CLEAN AIR FOR THE WHOLE HOME

Adequate filtration means controlling air quality as it enters the home and cleaning the air as it recirculates. But surprisingly, standard practice in HVAC installation is to route fresh air directly into the home without filtering it first.

Whole-home air purification systems utilize filters that must be changed periodically. A home's HVAC system has its own filters that should be changed frequently as well. Filters that are rated MERV 13 or higher are preferable. Some disposable filters even contain charcoal, which absorbs odors well.

GEEK BOX | the power of MERV

A filter's MERV rating measures its ability to capture small particles ranging from 0.3 to 1.0 micron. The higher the rating, the smaller the particles the filter can catch. Ratings range from 1 to 20, but the ideal range for a healthy home is 13 and up.

Keep in mind that if you go too high, your HVAC system will have to work harder because it's pulling air through a finer filter. This can raise your energy bills while lowering the unit's life expectancy. Thirteen is the sweet spot; it provides strong filtration without punishing your HVAC system. (Typically, filters rated at MERV 8 or less are not effective at catching the smaller particulates that get lodged in human lungs and cannot be removed by the body, while a MERV 13 can filter out pollen, dust mites, carpet fibers, mold spores, bacteria, pet dander, tobacco smoke, and more. That said, it will still struggle with fumes or gases, which is why it's so important to build healthy from the ground up.)

Although any filter with a high enough MERV rating will produce air filtered at the same quality level, the material in the filter matters. Some materials last longer than others, and some cost less than others. For example, fiberglass is a common filter material due to its low cost and ready availability, but it's not especially good at capturing smaller particulates. HEPA filters, however, have a MERV rating of 16 or higher and remove 99.97 percent of airborne particulates as small as 0.3 micron, cleaning air well and often lasting a long time. Also pay attention to whether the filters are assembled using adhesives or mechanical fasteners.

Energy recovery systems (ERV) and heat recovery systems (HRV) are great additions to help keep clean air circulating throughout a home. These handy systems also ensure the home doesn't develop negative air pressure from the use of bathroom and/or kitchen exhaust vents. Negative pressure occurs when the pressure outside exceeds the pressure inside the home, which can cause air to be pulled from unwanted areas as the house attempts to equalize the pressure differential.

Be aware that ERV and HRV systems tend to produce condensation inside the unit at the warm and cool air exchange. As such, we advise against using these units in climates where exterior humidity often exceeds 60 percent.

Moisture control is crucial to a healthy indoor living environment. Any house will benefit from a whole-house dehumidification system. This system ties directly into the ducting and runs independently of the HVAC system, kicking on when needed and reducing your heating and cooling bills. Be sure to keep your dehumidifier clean (mildew free) and running smoothly with proper maintenance. More on these in the next chapter.

Ventilating bathrooms and laundry rooms is also critical for maintaining proper moisture control. Installing multiple moisture-sensing exhaust fans is a good idea to ensure humidity levels don't climb in these vulnerable areas.

In all our design-build homes, we pay close attention to window placement. First, opening windows and screened doors on nice days to let in fresh air is essential to a healthy home. We boost this effect through natural cross-ventilation. Second, it's important to allow natural light into all areas of a home, including walk-in closets and hallways if possible.

Ensuring the garage is well ventilated is particularly essential. Children have been poisoned because of car exhaust, gasoline fumes, and other chemical fumes penetrating garage walls and entering adjacent bedrooms.

The Benefits of Fresh Air with Cross-Ventilation

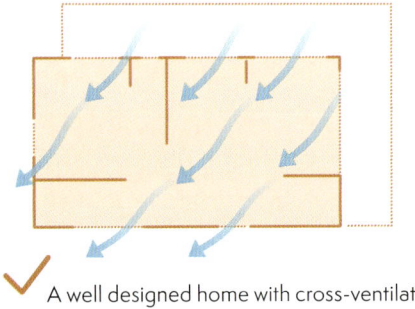

A well designed home with cross-ventilation

A poorly designed home that will have hot, stagnant areas

NATURAL GAS AND PROPANE

Natural gas is a popular and often economical way to heat homes in certain regions. A direct-vent gas heater is safe if the heating unit is not inside the home's conditioned envelope. We suggest planning for a mechanical closet that is accessed from the exterior of the house. Depending on the climate, the same closet may be used to house a water heater powered by natural gas or propane.

Although culinary enthusiasts often prefer cooking on a gas range, we advise against using natural gas or propane in the kitchen. Gas and carbon monoxide fumes are dangerous, and chronic low-level exposure has profound negative effects on the human central nervous system. In fact, some states are moving toward banning the sale of indoor gas appliances for this very reason. We say to stick with electric or induction cooktops.

Even with the right cooktops, effective kitchen exhaust systems are often overlooked. Never opt for a system that filters and recycles air back into the home. Stovetop exhaust should be routed to exit the home. Be sure to scale your kitchen exhaust system according to the size of your stovetop and the distance from the top to the vent fans. We encourage you to check out the Healthy Kitchens eGuide on the Healthier Homes website, which includes helpful kitchen hacks and valuable information about keeping a nontoxic, safe, and functional kitchen.

Homeowners often ask us about the safety of other gas or propane appliances, such as fireplaces. Select manufacturers do make direct-vent fireplaces that are completely airtight and vent outside the house.

A FEW WORDS ON DISHWARE AND COOKWARE

Kitchen items can be a source of air pollution in a home and may leach chemical residues into food. Avoid melamine, a chemical used in popular types of hard plastic dishware, utensils, and more. It's durable but cheap and can leach dangerous chemicals when heated or used with acidic foods. The same goes for items that contain bisphenols such as BPA, BPS, and BPF. Exercise caution with anything labeled "BPA Free," because who knows what the manufacturer used instead?

Glass, porcelain ceramic, stoneware, and stainless steel usually make great alternatives. But even with these materials, you need to be careful about lead and cadmium glazes. Both are toxic heavy metals. Even low-level exposure can cause severe harm, and children are especially vulnerable.

We prefer materials like silicone, stainless steel, lead-free glass, stoneware, and enameled porcelain ceramics. When you're shopping, make sure it's lead and cadmium free. Usually, the manufacturer's website will state this information.

On par with BPA are nonstick compounds like Teflon. Also known by the chemicals they contain, such as polytetrafluoroethylene (PTFE) and perfluorooctanoic acid (PFOA), these chemical coatings are made of fluoropolymers and can emit fumes and leach dangerous substances every time they're heated. Stay away. Other types of cookware that may leach include aluminum and copper.

In terms of safer options for pots and pans, you can use high-quality stainless steel (cheaper stainless alloys may leach, so look for options without metals like nickel), ceramic-coated cookware (again, with a lead- or cadmium-free glaze—look for 100 percent ceramic), and cast iron (make sure to brush up on how to care for it—for example, never use soap to clean cast iron, as that can ruin the seasoning).

A MATERIAL DIFFERENCE

As we've mentioned, the materials used in and around your home can affect the quality of your indoor air. Installing hard, nonporous surfaces that are easy to clean, such as tile or wood floors rather than carpet, will dramatically reduce the buildup of dust and dirt.

Dust and dirt that collect around baseboards, on fan blades, under beds, and in carpet can contain large amounts of allergens and mold spores. A bathroom with a continually wet, dusty corner will eventually turn into a mold hot spot. Thus, frequent vacuuming and dusting are essential to maintaining clean indoor air.

Using nontoxic HVAC components is just as critical for clean air as choosing the right HVAC system. HVAC components that perform well should neither off-gas nor negatively affect indoor air quality. That's because filtration and purification systems can do only so much. The best strategy for high-quality indoor air is controlling what goes into the home. Every product we use is specifically selected because it doesn't emit unhealthy fumes. We've researched and tested thousands of products and come to our own conclusions, regardless of what the products' marketing says.

For a healthy home, we use metal ducts or formaldehyde-free flex ducts, depending on the homeowners' budget. Unlined metal ducting is resistant to microbial growth, has less air-flow friction, and is easier to clean. One drawback is cost, which is where flex ducts shine. Before installation, be sure that metal ducts are cleaned with a power washer using nontoxic, industrial-strength detergent and warm water to remove any oily residues from the machining process.

GETTING COMFORTABLE

Thermal comfort for the occupants of a home is the end goal of any HVAC system, and comfort is influenced by multiple systems working in concert. That's why we recommend zoning sections of the home. For example, if the A/C is struggling to keep up with cooling the side of the house that faces the afternoon sun during the summer, zoning that area to run independently will prevent freezing out the bedrooms on the other side of the house. It's also beneficial to zone spaces according to their use, such as grouping bedrooms into one zone.

Thermal comfort can also mean energy savings, such as if you position the home to take advantage of the heat provided by the western sun's solar gain in the winter months. Most architectural CAD programs can do this.

THE ORCHESTRA

If one HVAC component is lacking, the other systems will inevitably suffer. Look at the example of humidity control. A lack of humidity control will cause condensation buildup on cooling coils and other metal components, which can lead to mold issues. Even if you have great filtration in this scenario, it won't stop the condensation and mildew problems. Mold can still spread through the ductwork and be blown into the living spaces.

The same can be said for sizing your system appropriately for the space and ensuring the duct sizes and registers can move the amount of air needed to effectively condition your home. We go into much more detail on the mechanics and sizing considerations in the Clean Air 101 eGuide on HealthierHomes.com.

The takeaway here is to size and install your HVAC system the right way, the first time, ideally at the beginning of a build project.

key points

▸ Indoor air quality is important but faces increasing challenges as building codes require homes to be airtight.

▸ Safe air purification systems are available, from HEPA filters to ozone generators.

▸ Install dampers to gain control over how much exterior air enters your home and keep your fresh air filtered.

▸ Pay attention to the placement of exterior fresh air intakes and exhaust vents.

▸ If you live in a dry climate, consider ERV and HRV systems to keep air circulating.

▸ To avoid compromising indoor air quality, we typically recommend against natural gas appliances within the building envelope, with a few exceptions.

▸ Using nontoxic materials is just as critical as choosing the right HVAC system.

▸ All these systems work together; if you skimp on one, you skimp on all.

Up next, let's talk water. ⟩⟩⟩

water

As builders, we try to think like water. That means carefully managing water from both interior and exterior perspectives and controlling its paths into and out of a home.

It takes only a few inches of rushing water across a road to lift a 2,000-pound truck and carry it downstream. Given the extreme force of water, imagine what a home endures when it rains. Water has tremendous power.

3 Inches of Rain in 1 Hour

For example, consider a home with a 10,000-square-foot roof area. Three inches of rain will produce over 300 gallons per minute (that's over 18,000 gallons per hour). That's the liquid equivalent of more than two semitruck tanker loads. Properly diverting that much water every minute requires careful consideration, even before breaking ground.

From the pitch of the roof to the grade of the land, your home's exterior water control systems must work harmoniously to ensure the inside of the home stays dry. Thorough planning during the early part of the design phase is integral to getting water control right the first time. We already covered exterior water control in previous chapters, so now we'll focus on water management inside the home.

START WITH QUALITY

First and foremost, good water quality is a key to healthy living. Whether your home draws from municipal sources or a well, we encourage you to perform a water quality test to help you determine what kind of purification system is best. Use caution when getting a water quality test from a local hardware store, as this can be a marketing tool to sell you one of the store's systems (which is OK if that's the one you want to purchase).

Some systems are installed at the point of use, such as filters on kitchen sink faucets or showerheads. These can reduce heavy metals, biological contaminants, carbon, pesticides, chlorine, petroleum products, industrial by-products, radioactive isotopes, pharmaceuticals, sediment, and other common water pollutants. However, since even low levels of exposure to these contaminants can cause a host of serious medical conditions, we recommend installing a whole-home water purification system at the point where the water main enters the house. Point-of-use systems tend to be less robust, and installing one at the kitchen sink, for example, doesn't help with the water in the bathrooms, for example. Think about the water in your shower, which you're not only touching with your skin but also inhaling as steam. A whole-home system is a simple, elegant solution that eliminates worry. If you have an older home, you may want to consider adding point-of-use filters in addition to a whole-home system due to the leaching potential of older pipes.

Cities are required to check water purity levels periodically and disseminate this information to the public. However, this system is far from perfect. Spot checks don't always catch periodic spikes in dangerous levels of water pollutants, and the EPA's list of acceptable levels seems to overlook the concern that even minute amounts of certain chemicals can be catastrophic to the health of people and animals.

LEAD AND OTHER CONTAMINANTS

Even at low levels, contaminants in water can have profound negative health effects. Worse, contamination is not only a problem in drinking water; many H_2O contaminants can be absorbed through the lungs and skin as well.

Take lead, for example. The EPA and CDC agree that lead is harmful in any amount, especially to children. It is not only toxic but also persistent in that it accumulates in the body over time. In adults, lead exposure can impair cardiovascular and reproductive systems. In children, it can cause behavior and learning problems, slowed growth, anemia, and other serious effects. Yet almost all municipal water supplies have detectable levels of lead contamination. Even more concerning are the countless older homes that have lead in their pipes and faucets, which wouldn't show up in a municipal water report.

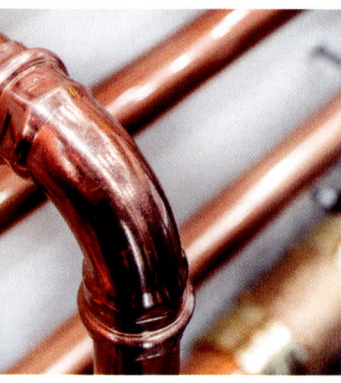

Other common contaminants include giardia, a tiny parasite that causes a variety of digestive issues; legionella, a bacterium that causes a type of pneumonia when inhaled or swallowed; and copper, a metal commonly used in pipes that, when it leaches into water in excessive amounts, can lead to numerous health problems.

GEEK BOX | the dangers of copper

Copper has been used to make water pipes since ancient times and has remained popular because it retains heat better than plastic. However, we avoid copper water supply pipes, as even industrial-grade copper can leach into the water supply.

We can absorb significant amounts of heavy metals through the skin, lungs, and stomach. Copper is no exception. Your body needs very small amounts of copper, but dangerous levels can creep up quickly and cause copper toxicity. Symptoms include headaches, fever, digestive issues, jaundice, and various mental and behavioral problems. Long-term copper exposure can damage the kidneys, liver, heart, or brain and even be fatal.

DECONTAMINATION METHODS

There are many water purification systems on the market, and they each have their pros and cons. The most effective systems for removing virtually all water contaminants and particulates are reverse osmosis (RO) systems, which use a partially permeable membrane to filter unwanted particles from drinking water. RO does have drawbacks, though. Whole-home RO systems are expensive, waste a lot of water and electricity during the filtering process, and require periodic maintenance by a water system professional. It's important to note that RO water is not the best for human or animal consumption and should not be used to water plants because of its lack of mineral content.

While it would be fantastic to have a one-size-fits-all water filter media, the reality is that different filter materials remove different contaminants. Contaminants vary by geographic region and the age of your utility system's underground infrastructure.

Whole-home water purification systems utilize a mix of different filters such as charcoal, activated carbon, ceramic or glass beads, and sand, and each kind is effective at absorbing or trapping specific contaminants per media. Filter treatments such as antimicrobial drip feeds, UV light, and ozone can be added to the system for antimicrobial purposes and to boost purity performance. You should mix and match these systems based on the results of your water quality test.

Water purification and softening systems are investments in your family's health. It's important to keep them sanitary, safe, and properly maintained. Don't forget to ask your water treatment professional how often the filters should be replaced.

ARE RO SYSTEMS WASTEFUL?

Unlike other water filtration systems, the reverse osmosis process uses between three and twenty-five gallons of water to produce a single gallon of RO water. (The amount varies based on the system and its condition.) During the highly effective filtration process, the water passes through a membrane and goes to a storage tank, while wastewater holding the contaminants drains. One idea is to install an outdoor holding tank and repurpose the wastewater to irrigate your lawn, but this can reintroduce contaminants into the ecosystem. Another drawback of RO systems is that they require electricity.

REVERSE OSMOSIS

| 1 gallon of RO water | = | 3–25 gallons of wastewater |

CLEANING THE POOL

A pool is a great place to hang out in the summer, but many cleaning treatments involve harsh chemicals. Here are some healthy alternatives:

- **Saltwater systems:** Although these systems do use chlorine, it's a natural process that uses drastically less by converting salt to chlorine gas to disinfect the water. Also called a salt cell system, it produces a gentler form of chlorine that's easier on the skin and requires less maintenance.

- **Ultraviolet systems:** UV light can kill cysts, algae, viruses, bacteria, protozoa, and even chlorine-resistant pathogens such as E. coli and giardia. It is completely safe and works in a variety of climates.

- **Ozone systems:** A powerful oxidant, ozone destroys microorganisms and stops deposits in water systems while requiring virtually no harsh chemicals. It is a newer technology and is becoming one of the most widely used water purification systems worldwide for a reason.

- **Magnesium mineral systems:** This new option is gaining traction in Australia and New Zealand. It utilizes a blend of magnesium chloride and potassium chloride, which are essential minerals for the body and make skin and hair oh-so-soft. The pool's filter uses glass beads and a salt cell system. We hope this fascinating technology soon becomes available in the US.

GETTING THE PIPES RIGHT

When it comes to plumbing, there are a few big no-no's to avoid. We do not run water supply lines through a slab foundation unless absolutely necessary, as a leak in the line could go undetected for years and result in extensive structural damage, flooring failure, and/or indoor humidity and mold issues. This is one reason why we recommend that homeowners avoid putting a sink or icemaker in a kitchen island. Interior drainage does run through the slab, but it's not under constant pressure like supply lines. We also recommend using PEX pipes instead of metal like copper or plastic made with PVC.

PEX pipes are known for their distinctive coloring, coming in red, white, blue, and gray, with each color denoting a different use. Red carries hot water, blue carries cold, and white and gray can do either. PEX is incredibly flexible and has great temperature resistance, but it can be damaged by sunlight, so it's best installed indoors.

GEEK
BOX **PEX pipes**

When it comes to delivering clean water to your faucets, toilets, and tubs, flexible PEX pipes offer a ton of advantages over PVC, traditional copper, and even galvanized steel pipes. They're lightweight, reliable, and easy to work with, meaning you can run water lines through walls without soldering or adhering a bunch of rigid pieces and elbows.

Controlling wastewater within a home starts with a sound sewage drainage system. Surprisingly, we've seen homes in which the lines were not installed with plumbing traps (P-traps), which are P-shaped pipes that contain sewage fumes and prevent them from entering the living space. That's a tough fix after sewer pipes are laid in a concrete foundation. This is one advantage of a pier-and-beam foundation, which does have accessible plumbing below the house. (*Note:* In the Foundation chapter, we establish why we do not recommend building on a pier-and-beam system unless building conditions require it.)

Drains should be accessible from the HVAC mechanical room, where A/C condensation collects and drains either outside the home or into a wastewater pipe. If condensation is routed directly outside the house, make sure the pipe is long enough that it doesn't constantly drip water onto the foundation. Creating a gravel drainage bed in this area is a good idea as well. Keep in mind that wasps or other insects can build nests inside drainage pipes, causing water to back up and leak into the home.

HUMIDITY AND CONDENSATION CONTROL

Humidity buildup and condensation inside a home are two often overlooked but equally important factors for interior water control. The Air chapter explains why condensation and humidity happen and how to keep them in check, while the Exterior chapter discusses how to prevent condensation problems around windows. Nonetheless, waterproofing is essential in places where interior water is supplied and/or drained, including kitchens, showers, wet bars, and laundry rooms.

Showers are the most problematic sources of leaks within a home. Using a nontoxic waterproofing system is an essential step to building out a shower or bathtub surround. (It's also great under sinks.) We run a contiguous polymer waterproofing membrane across the floor of every shower and all the way up the walls to the ceiling, which protects the wall cavity and directs water toward the drain if moisture gets behind the tile.

Many waterproofing applications used in showers are highly toxic, are subject to installation errors, and perform poorly compared to the waterproofing power of a polymer membrane system. We also source grout that contains no added solvents or formaldehyde. Larger tile is preferred in wet areas because it requires fewer grout lines.

Nontoxic silicone with no added petroleum solvents is your friend when waterproofing around toilets, sinks, shower glass, and so on. Don't forget to put a bead around decorative escutcheon plates and/or pipes that come through the tile. Store-bought construction silicones contain curing additives and fungicides that emit VOCs, even if the label says 100 percent silicone.

Water heaters are yet another place where water is stored inside a home. Newer tankless water heaters can be mounted on the outside of the house (depending on how cold your climate gets). This is a great option to mitigate the risk of indoor leaks should the heater fail. Plus, the gas versions do not require combustion within the home.

A water heater inside the home's envelope is OK if it is electric and has adequate measures built in to drain water from an unforeseen leak. An automatic water-sensing switch is important so that the system will shut off in the event of a leak. Some gas or propane water heaters can be installed via a direct vent system, which also keeps combustion and gas outside the home.

DEHUMIDIFIERS

A dedicated whole-home dehumidification system can be beneficial because HVAC systems are often the prime culprit when it comes to moisture problems. Mold grows more commonly and easily in HVAC systems than from leaks, and humans produce a surprising amount of moisture just from breathing and perspiring, so a dehumidifier may be needed depending on where you live.

A whole-home dehumidifier pulls moisture out of the air by chilling the air as it passes over coils cooled with refrigerant. As heat is removed, condensation develops. Then the system collects and drains the condensation, resulting in lower humidity levels.

We recommend a whole-home system that integrates directly into the HVAC rather than a stand-alone unit, which requires periodic emptying and will inevitably develop mold on the interior components. A whole-home dehumidifier, however, can be adjusted to ensure the perfect amount of dehumidification according to your home's dynamic indoor variables. It can even be hooked up to a thermostat that allows for independent control of the unit. That way, you don't need to have your HVAC system on for the dehumidifier to be running, which reduces the risk of a mold issue. (Never set your HVAC to fan mode, as it allows the fan to run without conditioning the air, resulting in condensation and mildew buildup.)

The only downside of a whole-home dehumidification system is the cost, especially if you're adding one to an existing home, which will require retrofitting and possibly adding space for the unit.

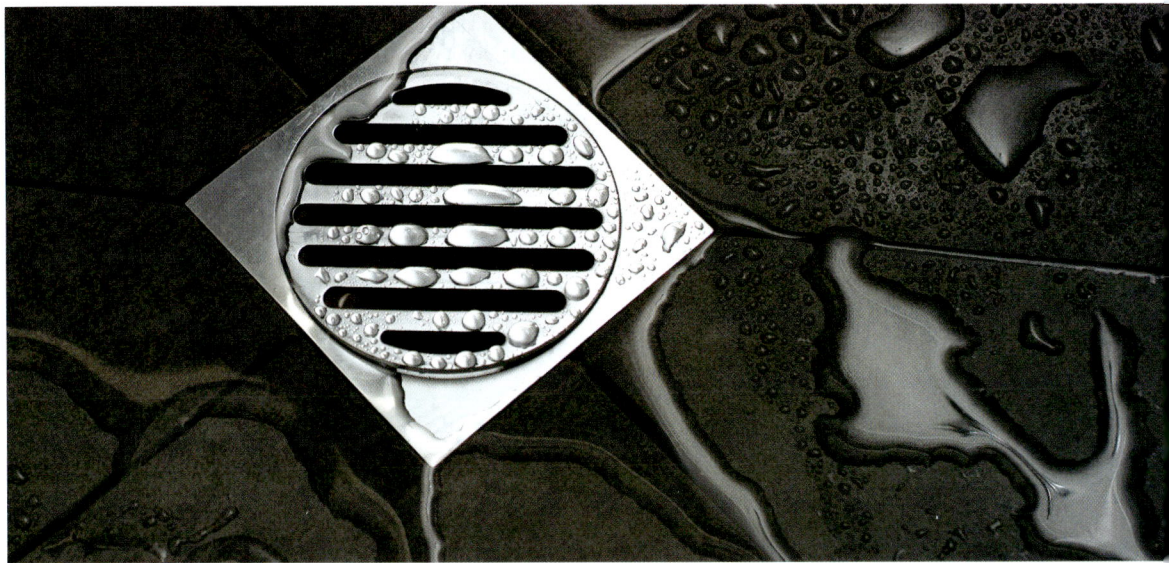

▶ Respect water. One rainstorm can dump thousands of gallons on a house.

▶ Managing water in a home requires looking at it from the inside out and controlling its means of entry and exit.

▶ Perform a water quality test to determine what kind of whole-home purification system your house needs.

▶ Be aware of lead and other common contaminants and choose a water decontamination system accordingly.

▶ Don't run water supply lines through a slab foundation if possible; any leaks can cause years of hard-to-find moisture issues.

▶ Use nontoxic waterproofing materials, especially in wet areas like bathrooms.

▶ Consider placing the water heater in an outside insulated closet or getting a tankless model.

Now let's head to the next chapter for information on all things energy.

energy

Although most forms of energy are invisible to the naked eye, there's no denying that energy plays a tangible role in our daily lives, meaning we must be aware of the role it plays in healthy building.

Industry and invention have brought about remarkable electronic discoveries that have transformed human life as we know it. Electricity powers our homes, letting us cook in our kitchens and enjoy one another's company in lighted rooms. We have access to hot water for showers, laundry, and dishwashing. Our homes are heated in the winter and cooled in the summer. Electricity equals comfort.

But the role of energy within our homes reaches far beyond electrical current, lighting, and warmth. Over the past several decades, we have introduced wireless technologies that emit other forms of energy along the EMF spectrum, such as radio frequencies (RFs) and microwaves. And, while certain forms of energy are beneficial to human life, others can be detrimental to our health.

WE'RE MOSTLY WATER

Think of our bodies as big bags of water—up to 75 percent of an average adult body is made up of water.

The human body operates using thousands of tiny electrical pulses generated within the central nervous system every second. At rest, humans produce around 100 watts of power and up to 2,000 watts in short bursts when sprinting, lifting, and so on.[1] This is possible because water is conductive, as are the electrolytes and minerals within us. We are electric beings, so it's no surprise that the energy and electricity around us can have profound effects on our bodies.

In this chapter, we will explain how to utilize electricity safely and reduce harmful EMF sources within an indoor living environment.

RESPECTING ENERGY

Remember that fun high school science class experiment that involved using hot dogs to conduct electricity? Rusty recalls using lemons rather than wieners, but the basic concept remains the same; electricity is a powerful phenomenon that comes with its own set of laws.

Contractors work with electricity daily, from designing functional electrical plans to determining voltage loads and strategically placing switches and plugs. We're especially cognizant of the local and regional energy codes and OSHA rules for job site safety. However, we take energy safety to the next level by also reducing the problematic health effects of chronic EMF exposure in a built environment.

Earlier in the book, we discussed energy considerations in the environment surrounding the home, such as avoiding powerful electrical fields from transmission lines that carry power over great distances. These tall structures create a much larger and stronger EMF than the service lines that wind through neighborhoods. It's important to consider the home's distance from exterior sources of frequencies along the electromagnetic spectrum. Some, such as sunlight, are beneficial in moderation, but chronic exposure to microwaves and very low-level EMFs can be detrimental to health.

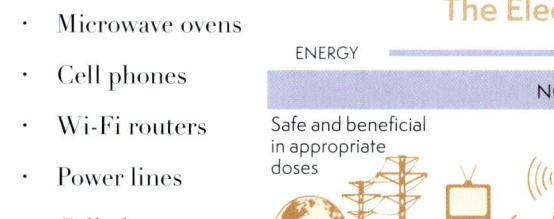

GEEK BOX
ionizing versus non-ionizing radiation

Before we get into securing your home against EMFs, let's address a common argument against taking such precautions: the debate about ionizing versus non-ionizing radiation.

Radiation exists all around us, and it takes one of two forms based on its wavelength along the electromagnetic spectrum. On the high-energy side of the spectrum is ionizing radiation, which can be very harmful depending on the dose. Even small doses can damage living tissue, eventually causing diseases such as cancer. Ionizing radiation can come from

· Outer space, such as gamma rays from the sun

· Terrestrial sources such as radon

· Man-made technology such as X-rays or CT scans

· Exposure to radioactive materials/minerals

On the low-energy side of the spectrum is non-ionizing radiation, which does not have the same harmful effects that ionizing radiation has on living tissue. It comes from

· Sunlight

· Microwave ovens

· Cell phones

· Wi-Fi routers

· Power lines

· Cell phone towers

· And more

The Electromagnetic Spectrum

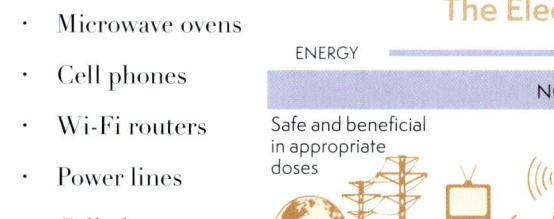

Because non-ionizing radiation does not alter but merely excites molecules—resulting in heat rather than severe and immediate cellular damage—some people and organizations do not view exposure as a long-term threat. But evidence continues to emerge that chronic exposure to certain types of non-ionizing radiation can cause cancer. For this reason, the World Health Organization has classified EMFs from cell phones and smart devices as potentially carcinogenic to humans.[2]

Much remains unknown about how non-ionizing radiation affects the human body. Until we can say definitively that these technologies are 100 percent safe, we prefer to err on the side of caution when it comes to smart wireless technology within the home. How about you?

We encourage you to look around your house. A good EMF meter comes in handy to determine just how far the field travels from electrical sources. Each power line is different, but you're looking for a reading of less than one milligauss (mG) inside the home's footprint. This is because the exposure in this setting is on a chronic and continuous basis.

If you're taking a reading within your existing home, you're still looking for a reading of less than 1 mG while not standing close to any appliances or wiring. We recommend turning on all the lights and appliances that would normally be operating while you are at home to get an accurate reading. If you don't live near power lines, a high reading may be due to improper grounding and imbalanced currents. Turn off the circuit breakers and take another reading to find out whether the problem is coming from the home's internal wiring or an outside source.

EMF SOURCES

Appliances have motors, and motors generate low-frequency EMFs when they are running. Such motors exist inside our refrigerators, washing machines, dryers, dishwashers, HVAC blowers, ERVs, and dehumidifiers. Motors are also present outside the home, such as in your outside HVAC unit and pool equipment. Breaker panels, multimedia centers, and wiring behind the walls are other sources of EMFs.

The good news is that electromagnetic field perimeters from motors and power lines drop off rather quickly as you move away from the source (as opposed to RFs and some microwaves, which are typically designed to travel medium to long distances).

Technologies that use RF frequencies are designed to emit in short bursts. These are older technologies like key fobs that require you to push a button to lock and unlock the doors.

Chronic microwave radiation within homes is a relatively recent phenomenon and a cause for concern. Microwave devices include Wi-Fi routers, Bluetooth-enabled devices, smartphones, smart home remotes, smart appliances, and devices such as wirelessly controlled light bulbs and locks. Amazon Fire TV sticks and Echo devices also utilize microwaves to communicate, as do some garage door openers, computers with Wi-Fi/Bluetooth antennas, and more. Home security systems and smart electric meters also often operate wirelessly. And the list keeps growing.

Many people don't realize that smart home features emit continuous microwave radiation unless you go into the menu of each device and disable the wireless connection. A handful of devices work the opposite way, where the wireless antenna must be enabled to connect and continuously emit wirelessly.

Avoiding connectivity inside the home is becoming an increasing challenge. We recently installed an oven featuring no warning that it emitted Wi-Fi frequencies that were impossible to disable. Certain cars and trucks now do this as well. Our newer model pickup allows the Bluetooth to be disabled, but only if the phone is plugged in via a USB cable.

While wireless technology offers its fair share of creature comforts and practicality, there's an unfortunate dark side to smart homes. Not only do these always-on microwaves present chronic health risks for people around the world,[3] but they are also a security threat.

No matter how hard security companies work to keep your information safe, hackers are constantly finding ways to break into wireless networks to steal personal data such as banking information and social security numbers. They can even intercept car key information to steal your vehicle from your driveway (which happened to Jen's father). Furthermore, these frequencies announce your presence in any given location, like a beacon.

TO SHIELD OR NOT TO SHIELD

Shielding is available for EMFs on RF and microwave frequencies. The important thing is to use the right kind of shielding materials for the frequency you're trying to block. Some shielding materials require a ground line to be run, either into the earth with a stake or into the ground through an electrical socket. There's even EMF-shielding paint that is quite effective at blocking specific frequencies in certain applications.

That said, shielding gets technical very fast and is not always practical. Shielding can become especially tricky when it creates a Faraday cage effect. As we mentioned previously, one of our homeowners wanted to use foil-backed plywood as the sheathing on the walls around her home. The goal was to block wireless frequencies coming from neighboring houses. After her home was built, however, we found that any type of device that emitted RF or microwave frequencies inside the home was amplified, as the frequencies would essentially bounce off the walls rather than exit the house. The same thing would be true inside an all-metal building.

The point is that shielding is a good idea in certain situations, but it can be tricky if not done properly.

Since shielding is costly and difficult to get right, our philosophy is to design and build homes so that motors are placed away from sleeping and hangout spaces. We place circuit panels far away from living areas, install whole-home surge protectors on the panel, and bury the wire from the service line so that it avoids running close to any bedrooms. Ideally, the power will come straight to the breaker panel without detouring around the home.

Electric company smart meters are placed on a pedestal away from the home, if possible. And we vet our HVAC components and appliances to ensure that any Wi-Fi and Bluetooth capabilities can be disabled and wired systems can be used instead. Furthermore, we hardwire smoke detectors, internet and cable connectivity, alarm systems, and A/V equipment.

Wiring a home to minimize electromagnetic disturbances is pretty straightforward. Keep the wires neat and tidy, avoid running them near places where people spend a lot of time, and stay away from adjacent construction materials that may be conductive.

Faraday cages

We've established that electricity is important to our modern lives but that too much can damage our bodies. It can also harm other modern electronics through interference.

Fortunately, nineteenth-century inventor Michael Faraday created the Faraday cage, a type of enclosure that shields its contents from electromagnetic fields. These cages can range from a metallic mesh to special sheet metal and even specialized paints. Regardless of their appearance, their purpose is to redirect electrostatic charges and electromagnetic radiation around the exterior surface.

By redistributing that energy, a Faraday cage deflects the radiation within its interior, acting as a hollow conductor that protects whatever's inside.

Faraday's concept also works by entrapping frequencies to keep an unwanted signal from penetrating a space. A prime example would be a small box with a metallic lining that traps the continuous signals from a car's RF key.

EMF Canopy Shielding

Continuously Emitting RF Keys

PLANNING FOR YOUR NEEDS

Always consider your energy needs when planning your build. For example, if you want an indoor sauna (which is a fabulous way to detox and improve circulation), be sure to plan accordingly. Some saunas require more amperage than others, so we always run a dedicated circuit for the sauna in addition to a ceiling exhaust fan to dissipate the heat.

LIGHTING YOUR HOME

Dimmer switches and compact fluorescent light bulbs (CFLs) often cause transient radiation, also called dirty electricity. The audible buzz that dimmers and CFLs produce is what transient radiation sounds like. Avoid these to reduce EMFs in your home. Incandescent bulbs and most newer LEDs made by reputable manufacturers are safe alternatives.

The color temperature of the bulbs used in a home also makes a difference in the look and feel of the space. Color temperature comes up on nearly every custom home project, and women almost always care more about the look and feel of ambient indoor light than men. In fact, studies show that women are generally able to distinguish between colors better than men. The traditional soft yellow cast of an incandescent bulb is best replicated by 2700K LEDs. Some people prefer a whiter natural light, which can be achieved with 3000K LEDs.

Ample lighting is especially important in places where tasks are carried out, such as kitchens and offices. After all, no one wants to slice a finger right before dinner. We often put each series of ceiling lights in the same room on its own switch so that the amount of light can be adjusted without using dimmers.

Abundant natural light in a home is essential to keeping your circadian rhythm on track. We design homes with no dark rooms or corners. While we do not install skylights in roof decks, as they inevitably leak, we incorporate natural light into every room possible, including larger closets and hallways.

While we're on the subject of natural light, solar panels are a popular option for harnessing sunlight as energy. They're cool but pricey, and you may not recover the expense unless your electric company will buy back the power you generate but do not use. The only EMF hazard with solar panels occurs where the power inverter is located, which is usually built into the panels. Although this can be shielded, it's easier to avoid the issue altogether by placing the panels over a garage or other area that is far from living spaces.

GEEK BOX

women have an eye for color

You might think men and women disagreeing about color choices is rooted in cultural conditioning, but research suggests the difference may be neurological.

Behavioral neuroscientists believe our brains' wiring affects our perception of color. Consider the results of a study by Israel Abramov, a behavioral neuroscientist at CUNY's Brooklyn College, who found that women are better at distinguishing between subtle gradations in hue, while men are better at perceiving changes in brightness across space, which is more useful in detecting rapid movement.

These results suggest that we are, in fact, wired differently for color. This may be because of hormones like testosterone, varying levels of which cause differences in the organization of neurons in the visual cortex. Men have more testosterone receptors than women, especially in the cerebral cortex's visual region, so this could affect visual perception.

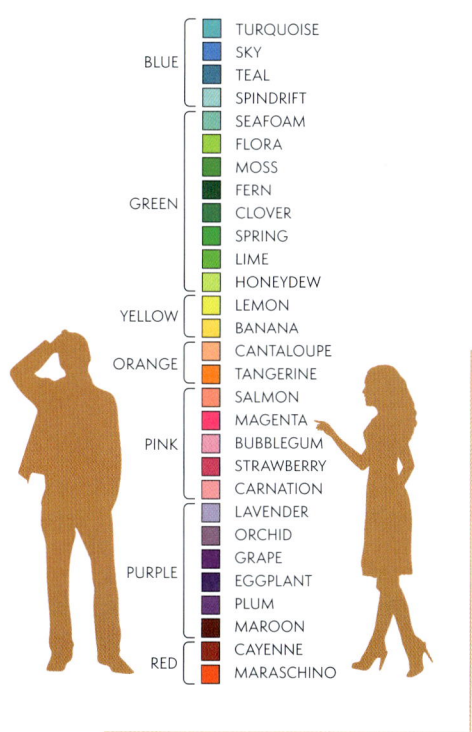

Women See More Color Gradients Than Men

BLUE — TURQUOISE, SKY, TEAL, SPINDRIFT

GREEN — SEAFOAM, FLORA, MOSS, FERN, CLOVER, SPRING, LIME, HONEYDEW

YELLOW — LEMON, BANANA

ORANGE — CANTALOUPE, TANGERINE

PINK — SALMON, MAGENTA, BUBBLEGUM, STRAWBERRY, CARNATION

PURPLE — LAVENDER, ORCHID, GRAPE, EGGPLANT, PLUM

RED — MAROON, CAYENNE, MARASCHINO

AVOIDING RADIATION

We've talked a lot about electromagnetic and microwave radiation, but what about ionizing (radioactive) radiation sources within a home? In the Foundation chapter, we discussed how to keep radon out of living spaces. We also mentioned that drywall made from synthetic gypsum is a potential source of radioactivity, which you can avoid by installing sheetrock made from natural materials. Did you know that we also avoid using granite in and around our homes, as it can sometimes be radioactive? This is why we do not install granite countertops (opting for other types of natural stone instead).

At the end of the day, energy is a beautiful thing. The laws of physics govern light, electricity, and all the frequencies along the electromagnetic spectrum. The best way to approach energy is to respect what it lets us do while remembering to work safely within its parameters.

key points

▸ Electricity powers amazing technologies, but we must respect it and mitigate the risks of EMFs in the home.

▸ Our bodies are mostly water, which means electromagnetic radiation—both ionizing and non-ionizing—can travel through our tissues and affect our bodies.

▸ Get an EMF meter to figure out the levels inside your home and fix hot spots.

▸ Be aware of EMF sources within the home—not only Wi-Fi routers and smart appliances, but also motors in machines like ceiling fans.

▸ EMF shielding should be considered but is not always practical and can create a Faraday cage effect, potentially amplifying signals within a home.

▸ Dimmer switches and CFLs can cause transient radiation; go with incandescent bulbs and newer LEDs instead.

▸ Be aware of potentially radioactive materials such as granite and avoid using them in your home.

▸ EMF considerations apply to various areas in the home, including heated flooring, electrical wiring appliances, fans, and more.

Keep reading for fit and finishing steps for your new home.

When contemplating building or remodeling a home, most people unknowingly skip straight to the fit and finish stage, which typically comes at the end of a construction project. Finishes are the elements inside the house that you can touch, see, feel, and use. They're like the icing on the cake—if the cake were your house.

So, we're nearing the completion of the project. The spaces of your new healthy house are built, and it's time to finish out the interior with style, color, texture, and character—to make it come alive, to make it yours, and to make it home.

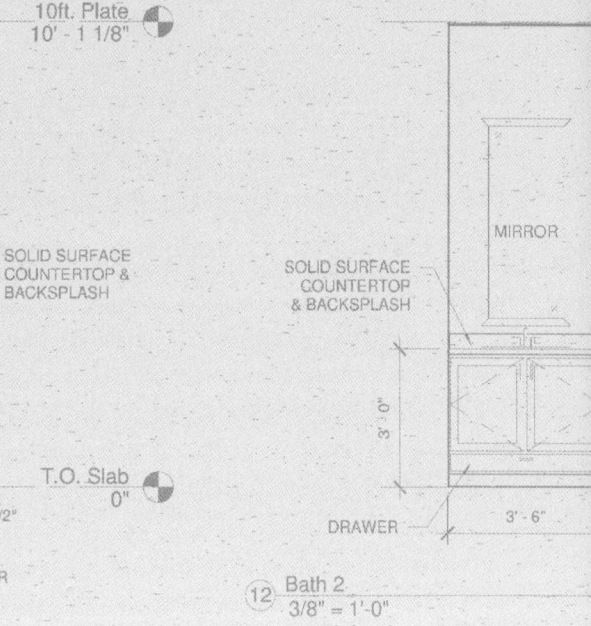

10ft. Plate
10' - 1 1/8"

SOLID SURFACE
COUNTERTOP &
BACKSPLASH

SOLID SURFACE
COUNTERTOP
& BACKSPLASH

MIRROR

3' - 0"

T.O. Slab
0"

DRAWER

3" - 6"

12   Bath 2
3/8" = 1'-0"

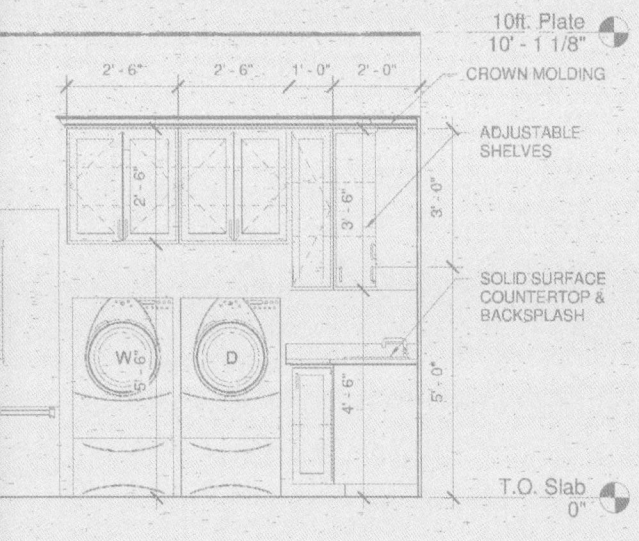

10ft. Plate
10' - 1 1/8"

2' - 6"   2' - 6"   1' - 0"   2' - 0"

CROWN MOLDING

ADJUSTABLE
SHELVES

2' - 6"   3' - 6"   3' - 0"

SOLID SURFACE
COUNTERTOP &
BACKSPLASH

W   D

4' - 6"   5' - 0"

T.O. Slab
0"

10ft. Plate
10' - 1 1/8"

2' - 6"

CROWN MO

ADJUSTAB
SHELVES

4' - 0"

MIRROR

OPEN

SOLID SURFACE
COUNTERTOP &
BACKSPLASH

4' - 0"

3' - 0"

T.O. Slab
0"

DRAWER

2' - 3"

7   Powder Room
3/8" = 1'-0"

10ft. Plate
10' - 1 1/8"

CROWN MOLDING

2' - 6"   2' - 3"   9' - 9"

10ft. Plate
10' - 1 1/8"
CROWN MOLDING
ADJUSTABLE SHELVES

4' - 0"

OPEN

4' - 0"

T.O. Slab
0"

10ft. Plate
10' - 1 1/8"
3' - 0"
CROWN MOLDING
ADJUSTABLE SHELVES

4' - 0"

OPEN

4' - 0"

T.O. Slab
0"

MIRROR

3' - 0"

DRAWER

2' - 8"

11 Bath 1 Water Closet
3/8" = 1'-0"

10 Bath 1 Vanity
3/8" = 1'-0"

PART 4:

finishing up

10ft. Plate
10' - 1 1/8"
CROWN MOLDING

ADJUSTABLE SHELVES

3' - 6"

SOLID SURFACE COUNTERTOP & BACKSSPLASH

1' - 6"

3' - 0"

T.O. Slab
0"

2' - 6"   2' - 9"

6 Pantry
3/8" = 1'-0"

10ft. Plate
10' - 1 1/8"

SOLID SURFACE COUNTERTOP

3' - 0"

T.O. Slab
0"

2' - 9"   2' - 6"   2' - 9"

5 Island Elevation 2
3/8" = 1'-0"

10ft. Plate
10' - 1 1/8"

2' - 8"   2' - 1"

2' - 0"   1' - 0"   2' - 9"   3' - 3"   3' - 4"   3' - 3"   2"   3' - 2 1/2"

finishes for walls, floors + ceilings

Eye-catching design and the functionality of the fit and finish inside a home are fundamental principles of architecture. It's like weighing pretty versus useful—these concepts encompass the previously discussed notions of form and function.

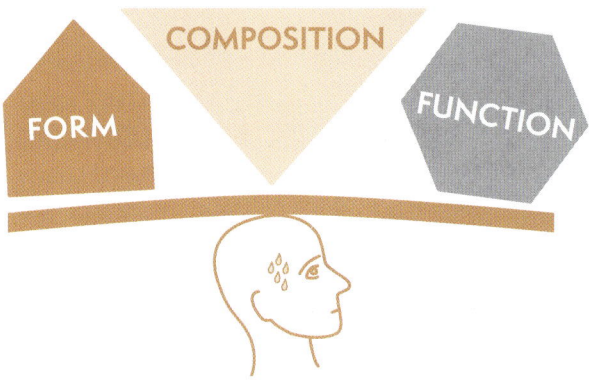

FORM IS THE STRUCTURE, FUNCTION IS THE OBJECTIVE

Paint, hardware, trim, lighting, flooring, tile, countertops—these are all finishes. As we stated earlier in the book, the relationship between form and function applies to the entire building process. Even though installing finishes (form) naturally comes toward the end of a build, it is crucial to plan for them at the beginning of any construction project. For example, you must plan for where countertops will go so walls can be placed and cabinets can be designed around the important working spaces that countertops provide (function). Remember, form should always follow function.

But we added a novel third component to our healthy building model. Our homes incorporate not only form and function but also healthy material *composition*. We introduced this fundamental element in Chapter 4, but composition takes another leap forward in our fit and finish processes. A home should not only be beautiful and functional but also provide a healthy and safe environment that fosters well-being, rest, and productivity. We achieve this by thoroughly vetting the composition of all construction materials and finishes that go into every one of our homes.

That said, there are a million choices for finishing a home, so adding the third element of composition to the decision-making process can quickly become overwhelming for builders and homeowners alike. In this part of the book, we lay out a framework of considerations that apply to each type of finish to ensure you feel confident in your selections. After all, a home can only promote wellness if it's built with composition in mind. From what is placed behind the walls to the final touches—it's all equally important in a healthy home.

FLOORING

Let's work from the ground up. Our favorite flooring is porcelain tile because of its durability, ease of cleaning, and low off-gassing potential.

Porcelain is a nonporous ceramic material made of minerals and clay that hardens when fired in a kiln at temperatures up to 2,600 degrees Fahrenheit. It's an excellent choice for wet areas such as showers and bathroom floors. Many porcelain tiles are glazed, meaning a layer of "liquid glass" is applied to the top, which allows for patterns such as wood or marble to be printed on the tiles. The result often looks astonishingly similar to real wood or stone.

Some homeowners have expressed concern about porcelain tile containing high concentrations of lead, but the truth is that most reputable modern tile manufacturers do not include lead in their porcelain or glaze. We always ask our tile installers to cut tiles outdoors with a wet saw to reduce the risk of distributing tile dust inside a home.

It's important to float even slightly uneven floors with nontoxic floor leveler to get longevity out of the tile you choose to install, too.

A Typical Healthy Home Selection Process for a Family with Young Children and Pets

COMPOSITION	FUNCTION	RULES OUT:	FORM	DECIDING FACTOR
HEALTHY CHOICES:	MUST-HAVES:	RULES OUT:	STYLE + COLOR:	QUICK INSTALL:
Concrete	Easy to clean	Concrete	Modern farmhouse	Click together
Tile	Scratch-resistant	Tile	Neutral colors	Wood look
Wood	Low-maintenance	Wood	Bohemian	No-VOC cork
Cork		Cork		DONE!
Carpet		Carpet		

the art of floating floors

When we describe a "floating subfloor," we mean a floor that doesn't touch the concrete foundation but instead rests atop an additional layer of specially poured concrete that levels out during curing via the force of gravity. This provides a level surface for the tile installers and ensures your flooring will remain secure for many years to come.

For flooring other than tile, which is placed directly on the foundation slab, the use of a vapor barrier is usually advised. The underlayment can be made of woven nontoxic materials, polyethylene, or even certain adhesives that have vapor-blocking properties. Just be sure to do your homework on the product ingredients and material compatibility. Hiring a professional flooring subcontractor is highly advised.

STONE, WOOD, CARPET, AND MORE

Keep in mind that although brick and stone are timeless and beautiful, these materials are quite porous and tend to chip or crack when used as flooring. A generous application of a nontoxic stone sealer is required periodically.

Wood flooring is also a great alternative, but you'll need to do some research first. Although expensive, solid wood plank flooring is a beautiful option. We recommend using an unfinished hardwood so that your painter can finish the planks with nontoxic products. Remember that hardwood floors should acclimate in a conditioned space before being installed.

We have found a handful of engineered wood flooring companies that meet our standards for zero added formaldehyde and zero VOC emissions. When shopping for engineered wood floors, remember to

- Ask which types of glues were used in manufacturing the planks. Stay away from adhesives that contain formaldehyde. Some manufacturers now use soy-based glues with no added formaldehyde or solvents.

- Avoid products with a synthetic rubber backing.

- Look for products with polyethylene terephthalate (PET) water-resistant topcoats (instead of PVC).

- Ask about click versus glue-down application options.

Luxury vinyl flooring (LVL), an attractive and affordable option, is all the rage these days, but almost all LVL flooring contains PVC (a plastic that will off-gas fumes for a long time) and resins that likely contain formaldehyde.

Cork is another great flooring option and comes in countless colors and patterns, including some that look remarkably similar to wood. The perks of cork include noise dampening and natural resistance to moisture and microbes.[1] When shopping for cork flooring, consider the same points previously listed for engineered wood flooring.

Carpet can be a blessing and a curse. It is soft and cuts down on noise. In fact, we're not opposed to carpet being used in dry places like bedrooms and home theaters. However, it tends to harbor dust, dirt, pet dander, and other allergens that are not conducive to a clean indoor environment. A handful of eco-conscious carpet manufacturers have put a lot of thought into creating natural nontoxic options. Look for natural fiber or wool carpet colored with plant-based dyes.

That said, many manufacturers apply dangerous stain repellents and fire retardants to carpets where babies, toddlers, and pets play. A thorough review of the current scientific literature suggests that caution should be exercised with carpet use, as it may cause allergies and asthma, especially in children.[2] Most carpets are also chock-full of formaldehyde and a laundry list of other harmful chemicals. Make sure your carpet manufacturer uses a backing such as natural latex and yarns with no-VOC adhesives. The best underlayments are wool or inert materials that are not recycled and do not have antimicrobial or fire-retardant coatings.

Concrete floors are an excellent, easy-to-clean option. Make sure you stain and seal or paint your floors with nontoxic products designed for concrete. Do not use epoxies inside the home. Another maintenance coat is required every few years to avoid the potential for spill stains. While concrete does show scratches easily, these imperfections can give the floor character. Concrete is hard and can amplify noise inside the home, but area rugs are a simple solution.

Be sure to discuss flooring options with your builder at the beginning of your construction project. Correctly prepping the home for flooring is largely dependent on the flooring thickness, weight load, and product installation requirements.

WALLS AND CEILINGS: SOURCING HEALTHIER PAINTS

The largest surface areas inside your home are the walls and ceilings, and getting them right is key to a comfortable living space with high-quality indoor air. A handful of big-name paint companies are now promoting low- and no-VOC interior paint products. This might seem great, but most of these big paint manufacturers still include ingredients such as fungicides, biocides, ammonia, acetone, paraffin, formaldehyde, and other chemicals that will inevitably release toxic fumes into your home. These chemicals aid in performance, providing faster drying times, better freeze-thaw resistance, and so on.

The EPA's list of VOCs is very short compared to the thousands of chemicals on the market today that are harmful to human and pet health. Many chemical solvents are considered VOC-exempt by the EPA for use in manufacturing paint, adhesives, inks, and coatings simply because they don't harm the Earth's ozone layer.

One example of the many exempt solvents is tert-butyl acetate,[3] which was recently granted exempt status from the VOC and HAP lists but is included in the production of lacquers, enamels, inks, adhesives, thinners, and industrial cleaners (all items that are heavily used in conventional home construction). This chemical is also listed on the New Jersey Department of Health's Hazardous Substances List and is cited as irritating to the eyes, lungs, throat, and skin. Tert-butyl acetate is a skin sensitizer, and if an allergic reaction develops, even low future exposures may cause itching, rashes, dryness, and cracking for years to come. This chemical has yet to be tested to see if it causes reproductive harm or is a carcinogen. It

VOCs VERSUS HAPs

Hazardous air pollutants (HAPs) are the EPA's label for pollutants in the air that may be harmful to humans and animals. These differ from VOCs (and are a bit more accurate for our purposes) because not all VOCs are toxic. But, since we're talking exclusively about how to create a healthy home environment (which means the goal is to eliminate all airborne pollutants), we may sometimes use the terms interchangeably. Bear in mind that all Healthier Homes construction products are rated zero for HAPs as well.

was granted exempt status on both lists in order to offer an alternative to xylene, another example of a "regrettable substitution." Who wants this stuff all over their walls?

By now, you know that labels on construction products can't always be trusted. You also know that nothing we use inside a healthy home is derived from off-gassing petroleum materials. Some paints touted as natural and clean contain petroleum-based waxes, and just because a paint claims it's water-based doesn't mean it's a healthy choice to put on your walls.

A savvy builder and an educated homeowner will do their homework when sourcing paint, primer, stain, and clear coat. The best kinds of paints and topcoats are truly zero VOC and act as a barrier to seal any potential off-gassing from its substrate. We typically avoid oil-based paints, which off-gas, and latex paints, which lack durability. Instead, we recommend water-based acrylic paint, which tends to be durable, easy to use, and truly low- or no-VOC. You may pay a little more for this type of paint, but in our opinion, it's worth it.

We mentioned that paint covers the largest surface area of your home. A challenge we've continued to face is the limited availability of truly nontoxic, well-performing paint options. This is why we are introducing our own line of zero-VOC Healthier Homes paint and construction products that we trust and use in the homes we build.

5 TIPS FOR SOURCING HEALTHY PAINTS

Know your VOCs: Learning the most common sources of VOCs and HAPs found in paints will help you keep an eye out for paints to avoid. Some of the most common include toluene, xylene, methyl ethyl ketone, and ethyl acetate.

Learn the labels: As discussed elsewhere in this book, labels can be misleading. Remember that "low-VOC" paints typically contain fewer than 50 grams of VOCs per liter, while "zero-VOC" is less than 5 grams per liter. Doesn't sound like "zero" to us!

Keep an eye on the base: The base of a paint is often the main source of VOCs, so make sure to double-check that before mixing. Water-based ones tend to have lower VOC content than oil-based options.

Watch out for certs: As we've mentioned, certifications aren't everything. They can sometimes help you identify less risky products, but beneficial certifications like GREENGUARD and LEED still allow low levels of some harmful VOCs and HAPs.

Always dig deeper: Some manufacturers are trying to be more transparent about their ingredients and practices nowadays, but it's nowhere near the majority. That means it's still your responsibility to find out what's in your paint before buying.

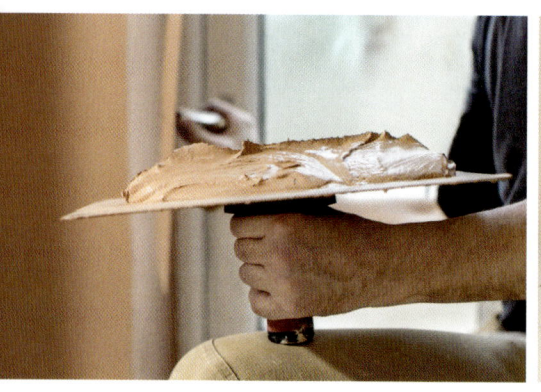

PLASTER, SHIPLAP, AND MORE

Plaster is another option for walls. Healthy clay-based plasters produce colorful results. Even though the manufacturer says this plaster is OK to use in bathrooms, we advise against its use in wet areas due to its absorbent nature. Some people also do not like the slightly earthy mineral scent of the clay, so try it in a small area first to test for tolerance. Be aware that plastering has become something of a lost art, so it may be difficult to find a skilled artisan who is familiar with clay plaster installation. Stay away from plaster products that are not clay-based.

Shiplap, a type of tongue-and-groove wooden board, is a great way to add texture to a wall or ceiling. After the TV series *Fixer Upper* brought back this trend from the previous century, many shiplap products have emerged. Look for unfinished tongue and groove made of solid wood.

Hard surfaces such as brick, stone, ceramic, porcelain, glass, and quartz can also be used for walls; the same considerations from the tile and stone sections apply. When installing large pieces, make sure the adhesive and/or thinset is rated for the weight and doesn't contain formaldehyde or petroleum distillates.

Wallpaper is back in style, but the adhesives are typically far from healthy. Moisture also tends to get trapped behind it over time, creating a breeding ground for mold. Avoid wallpaper in bathrooms, laundry rooms, or other areas prone to humidity, and look for papers made of natural materials such as cellulose with water-based inks and no vinyl coatings. The Healthier Homes online shop recently released a collection of woven wallpapers made from reeds and grasses, and we've formulated our own Healthier Homes adhesive. Dramatic alternatives include porcelain or ceramic tile, stone ledgers, and paint stenciling.

Ceilings can be treated mostly the same as walls. That said, avoid popcorn ceilings. Most kits no longer contain asbestos, but even the cleanest popcorn ceiling will accumulate dust faster than other finishes.

key points

▸ Although the fit and finish stage comes at the end of a construction project, it's important to plan for it at the beginning.

▸ Porcelain tile is our favorite flooring because of its durability, ease to keep clean, and low off-gassing potential.

▸ Wood, tile, stone, cork, and cement are all good flooring options worth careful consideration.

▸ The walls and ceilings are the largest surfaces in your home, so it's important to get them right.

▸ Don't trust the label on paint products: some big companies offer interior paints that claim to be low- and no-VOC but do off-gas unregulated VOCs.

▸ Certain woods, clay-based plasters, and other materials can work for walls, but be careful to choose HAP- and VOC-free stains, finishes, and adhesives.

For more on healthy interior finishing options, including paints, primers, and other coatings, check out the HealthierHomes.com store or continue to the next chapter for the lowdown on all your other interior surfaces.

smaller surfaces

Floors, walls, and ceilings may be the largest surfaces in a home, but that doesn't mean the others are undeserving of careful consideration as well. From a tiny nailhead to the expansive kitchen countertop, every surface counts inside a healthy home.

COUNTERTOPS

Countertops are workspaces that should be durable and easy to clean. They're also decorative, so beauty is important, too.

Our favorite solid surface for countertops is manufactured quartz. These slabs are made of natural quartz stone and inert resin, and we use them in all our home builds. The result is a durable, beautiful, and timeless countertop that mimics the look of other popular countertops such as marble, granite, slate, and concrete. Manufactured quartz will not off-gas and has a near-zero porosity, making it impervious to the bacteria that cause disease and trigger allergies.

We do not install granite countertops in our homes because some slabs have the potential to be radioactive. Other natural stone products such as marble, soapstone, slate, and quartzite make gorgeous alternatives. Keep in mind that they do tend to stain and require periodic maintenance.

Stay away from cultured stone or cultured marble (also known as faux marble), which can be made from unhealthy materials and lack durability.

air-purifying ceramics

An Italian countertop manufacturer has figured out how to incorporate photocatalytic technology into countertops to offer anti-bacterial, anti-pollution, and anti-odor properties. Created through an application of micrometric particles of titanium dioxide and silver, these high-quality porcelain slabs also boast self-cleaning properties by reducing dirt adherence.

Surfaces using the patented technology—which was developed in collaboration with the Chemistry Department of the University of Milan—create an active photocatalytic surface (which works even in the dark or under LED lights) that eliminates 99.99 percent of bacteria and viruses and cleans nitrous oxide from the air. Even better, these features last for the life of the slab.

Another European countertop manufacturer recently patented their own solid-surface countertop material that boasts antibacterial surface properties to create germ-free, hygienic, food-grade prep surfaces. Better yet, these countertops continuously purify the surrounding air and can even self-clean. Pretty neat technology!

Glass and metal are also great countertop options. Porcelain countertops are the newbie in the countertop world and are also a good option. Some manufacturers are even working on technology for nontoxic stone-look countertops that naturally purify the air using ambient water and light.

Make sure your countertop installer uses nontoxic adhesives during the glue-down process. If your installation includes a backsplash of the same material, use silicone of a similar color with no added petroleum distillates.

Avoid using MDF or particleboard for your countertop substrate. Other materials to stay away from include epoxies, surfaces with baked-on enamels, some acrylic polymer products, and laminated countertops.

HARDWARE

Plumbing fixtures, lighting, door handles, drawer pulls, and cabinet knobs are all hardware decisions. The considerations are fairly standard and straightforward. As with most purchases, you get what you pay for. Choose a reputable manufacturer that uses high-quality factory finishes on durable substrates.

Contrary to popular belief, there is no cause for concern about lead when it comes to purchasing new plumbing hardware, such as faucets and showerheads. In the US, federal law requires that manufacturers produce lead-free plumbing hardware (the legal limit is 0.25 percent). That said, beware of antique or second-hand hardware, as these could leach lead via touch and into the tap water.

Also stay away from genuine pewter and genuine brass. These metallic alloys can include lead and antimony, which can leach into water and be absorbed when you touch the metal.

BATHTUBS AND SHOWERS

Just like with kitchens, bathrooms will get you plenty of bang for your buck in terms of home value. High-quality fixtures and healthy materials are worth the investment.

Bathtubs now come in a wide variety of materials to make them lighter and less expensive. Acrylic tubs are durable and a great option in terms of cost and performance. Traditional iron tubs are also a good option, as they hold heat the longest. On the pricier end of the bathtub spectrum are the "solid surface" tubs made of stone, minerals, and acrylic resin. These are beautiful and durable and have a soft, natural look and feel.

Cheaper tubs are often made of low-quality fiberglass, as are prefab shower kits. We do not install these fiberglass products. They tend not to hold up over time, and the materials used in their construction can emit harmful VOCs. They are also subject to cracks after some wear and tear, and no one wants a leaky tub.

Showers can be one of the weakest links for water penetration into a home, so we take shower tile installation very seriously. A waterproof nontoxic membrane should run from floor to ceiling to capture and drain away water as an added measure of protection in case water somehow finds its way behind the tile.

CABINETS AND TRIM

The average home has more cabinets than you might think. We have to consider the kitchen, pantry, mudroom, laundry room, bathrooms, closets, living room built-ins, bars, garage storage areas, playrooms, and more. Cabinets are important for organization and storage and can have a giant impact on a home's indoor air quality. That's why, as innovators in building homes without harmful chemicals, we require cabinets to be constructed to exacting specifications.

Currently, there are two ways to purchase cabinets:

- Prefabricated from a big box store

- Custom-made from a local carpenter or high-end manufacturer

Let's start with the first option. Prefabricated cabinets are cheaply made, extremely toxic, and will off-gas for years. Made of particleboard, MDF, PVC, and laminate, these use cheap glues and oil-based finishes and are a major source of pollution in a home. You will never find prefab cabinets in any of our healthy homes.

The second option—hiring a local cabinet shop or specialty manufacturer—is by far the best way to source custom cabinets. It may take some time, but it's worth it to shop around and find a skilled carpenter or high-end custom outfit that is willing to build your cabinets with the materials and methods you want.

Our in-house carpenter does amazing work and is one of the most skilled woodworkers we've ever seen. However, his small shop couldn't keep up with the demand from our numerous construction jobs. (Remember that list of rooms requiring cabinets?)

This challenge led us to partner with a high-quality custom cabinet manufacturer that was willing to not only take on the workload but also create a custom line that meets our unique and stringent healthy cabinet specifications.

Recognizing that this is a real bottleneck with healthy home building, we're working to change the lack of availability of beautiful, quality-built, nontoxic cabinetry.

In the meantime, ask your local carpenter to

- Use virgin solid hardwood cabinet interiors and drawer box construction
- Use nontoxic wood glues and caulk that won't off-gas
- Source cabinet doors and drawer fronts made of solid hardwood
- Use your own zero-VOC paints and stains
- Install high-quality drawer slides and hinges—you'll be glad you did

The number of customization options for cabinets, baseboards, and window and door trim is truly endless. Small touches for cabinets like adjustable shelving, pull-outs for trash cans with integrated cutting boards, and organized dish storage are some of the cool options available. Order your cabinets with upgrades like soft-close dampers to eliminate slamming and increase the longevity of your cabinets. Even removable toe kicks are game changers for cabinet functionality.

One of our favorite alternatives for wall storage includes floating hardwood shelves and powder-coated metal rack systems. Floating shelves are a statement piece and are anchored to the wall from within, appearing to "float" with no supports. Stylish yet minimal stand-alone metal racks are an economical answer to kitchen storage woes and can go in just about any kitchen. Check out the chapter on sourcing healthy furnishings for more kitchen storage solutions!

APPLIANCES

When it comes to fits and finishes, the conceptual interior design process starts on day one of every project. To the surprise of our homeowners, we begin with appliance selections, as these dictate plumbing and spatial allocations for drafting the plans in critical areas like the kitchen and laundry room. (Plus, lead times for ordering appliances can be long, currently up to a year.)

You may not think of your oven as a statement piece, but appliances play a large role in setting the tone and mood of your home, and there are lots of color and configuration options. For example, refrigerator choices used to be limited to stainless steel, white, or black and upright versus side-by-side. Now, there are a hundred shades of dark stainless steel with more configuration options than your master closet.

   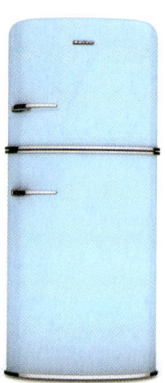  

Be aware that numerous appliances have wireless capabilities. If you prefer to avoid constant wireless transmission, check to see if the Wi-Fi or Bluetooth antennas can be disabled. If the transmitters can be turned off, there should be no issues with the appliance continuously emitting frequencies.

Motors inside appliances are also a source of electromagnetic radiation. Place your appliances away from sleeping areas or spots where you spend a lot of time (for example, your work desk or media room couch). Pay attention to the appliances directly above or below a living space as well. The reach of the EMFs emitted from these motors varies; however, they usually travel only a few inches and rarely extend beyond several feet.

We've seen instances where a bed shares a wall with the refrigerator in the kitchen or the washer and dryer in the laundry room. One of our homeowners recently realized that their son's bed shared a wall with the swimming pool equipment in their previous home. It was a light bulb moment for them, as the child had spent years dealing with sleep difficulties and would often move to the couch to rest. This constant exposure to electromagnetic radiation can disturb sleep patterns for many people and may lead to chronic sleep deprivation.

Induction cooktops are wonderful to cook with, as they create heat quickly and evenly. However, those who are sensitive to EMFs may be better off going with electric cooktops, as the EMFs are not as strong.

Microwaves are commonplace and do make heating food fast and easy. However, these handy devices are also intense sources of microwave energy and, in rare cases, can leak small amounts of radiation. We recommend opting for an oven with the air fry option instead, which cooks quickly and efficiently.

Also pay attention to what's inside your oven. Some self-cleaning ovens contain nonstick surfaces such as polytetrafluoroethylene (PTFE) coatings. Heating the oven to very high temperatures in the self-cleaning mode can release toxic fumes into the home. These fumes can kill small pets such as birds.

We recommend avoiding the use of natural gas and propane appliances inside the home. Natural gas–burning appliances are found in millions of US households, but new research suggests they may produce pollutants that are harmful to human health, from the commonly acknowledged carbon monoxide (CO) to the lesser-known nitrogen dioxide (NO_2). The negative health impacts of NO_2 are particularly alarming because they seem to be more pronounced in children, causing learning deficits, respiratory and cardiovascular problems, and more.[1] In fact, California is moving toward banning the sale of indoor natural gas appliances in the coming years.[2]

Direct-vent fireplaces are the exception to the no-gas rule in a healthy home. These systems are fully enclosed and vent outside the house.

| **key points** | ▶ Every surface counts inside a healthy home. |

▶ Every surface counts inside a healthy home.

▶ You have a lot of options for appliances. We recommend against gas as well as any device that emits strong or constant EMFs.

▶ Go with quartz, glass, porcelain, or metal countertops; avoid toxic adhesives and caulks, MDF or particleboard, and other emitting materials.

▶ For hardware, choose a reputable manufacturer that uses high-quality factory finishes that won't tarnish or break.

▶ Acrylic, solid surface stone, and iron bathtubs are good options; avoid fiberglass and prefab shower kits. Instead, go with ceramic or porcelain tile.

▶ Cabinets can be a huge source of VOCs; we recommend finding a skilled local carpenter or high-end manufacturer who can make them to your specifications.

Up next, our guide to sourcing nontoxic furnishings for your new healthy home.

Sourcing nontoxic furnishings is just as important as using the right construction materials. The easiest way to shop for furnishings, decor, and accessories for a home is online, where you have access to product descriptions and can easily look at the components of each piece of furniture. For example, a couch will basically consist of

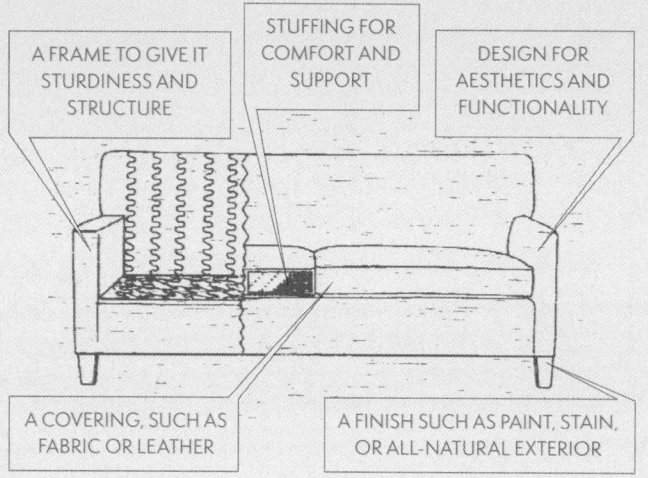

A FRAME TO GIVE IT STURDINESS AND STRUCTURE

STUFFING FOR COMFORT AND SUPPORT

DESIGN FOR AESTHETICS AND FUNCTIONALITY

A COVERING, SUCH AS FABRIC OR LEATHER

A FINISH SUCH AS PAINT, STAIN, OR ALL-NATURAL EXTERIOR

Sound familiar? The frame is the "form," just like the foundation, framing, walls, and roof serve as the structural form of a home. The stuffing, design, covering, and finish serve as the "function," which is to provide a relaxing and stylish place to sit.

Remember, the third component is "composition."

Sourcing healthy furnishings is all about taking a closer look at material composition, which is why this part of the book breaks down each of the major materials used in home furnishings, starting with wood.

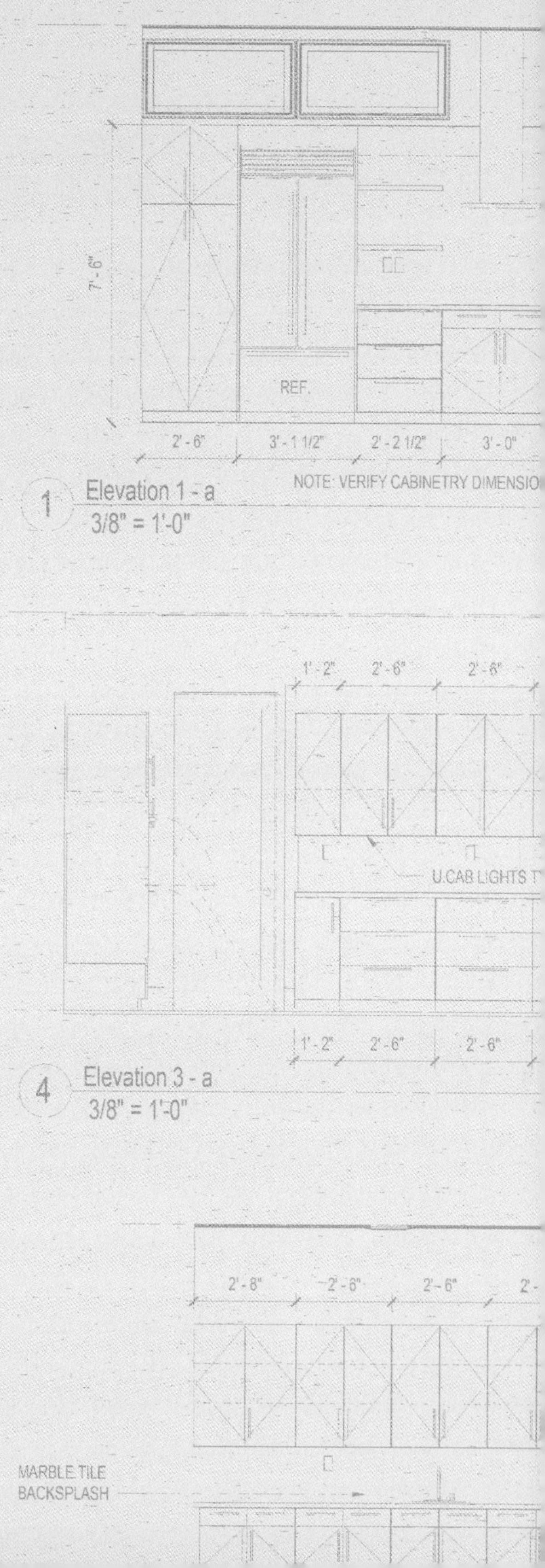

7'-6"

2'-6"   3'-1 1/2"   2'-2 1/2"   3'-0"

REF.

NOTE: VERIFY CABINETRY DIMENSIO

① Elevation 1 - a
3/8" = 1'-0"

1'-2"   2'-6"   2'-6"

U.CAB LIGHTS T

1'-2"   2'-6"   2'-6"

④ Elevation 3 - a
3/8" = 1'-0"

2'-8"   2'-6"   2'-6"   2'-

MARBLE TILE
BACKSPLASH

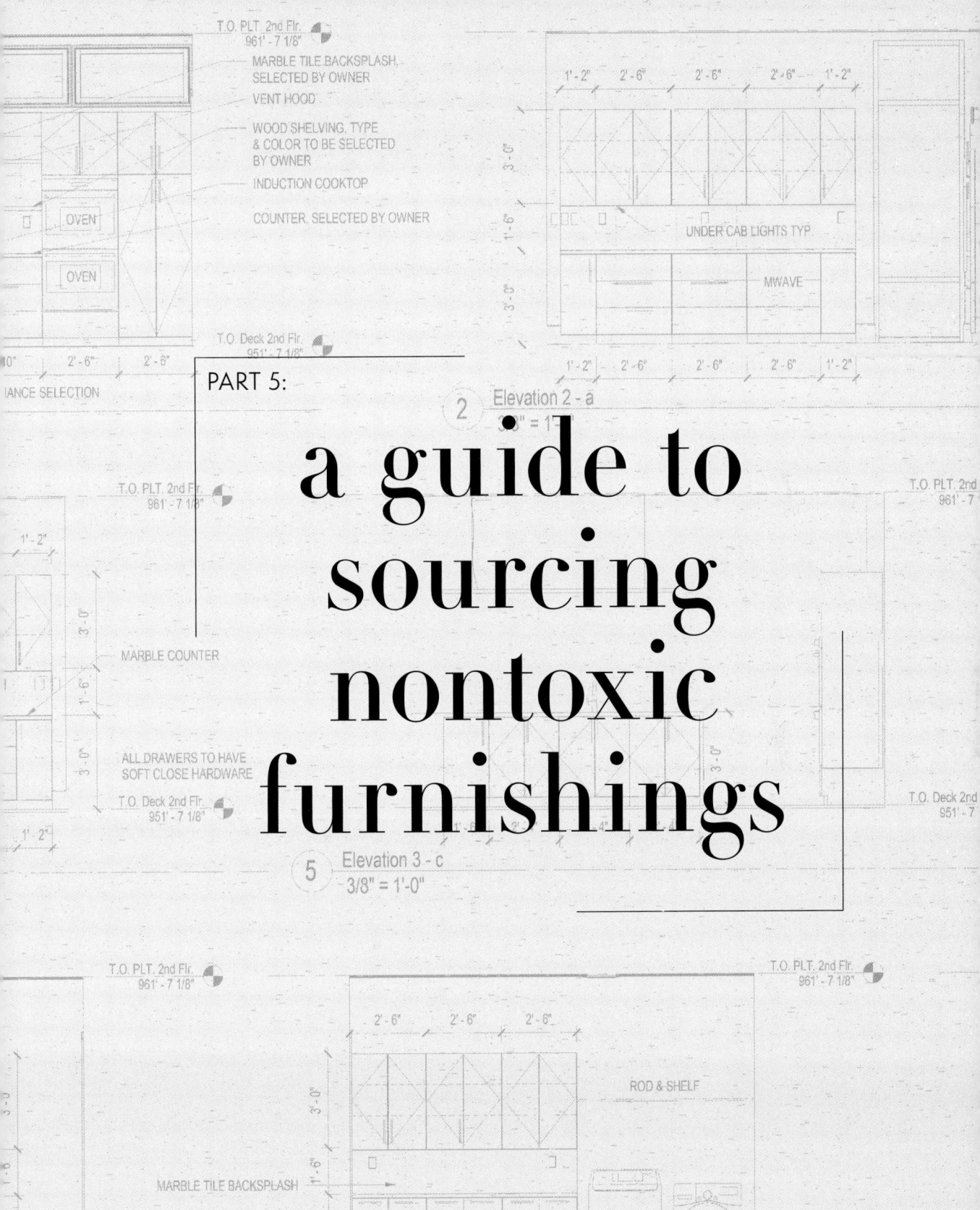

PART 5:

a guide to sourcing nontoxic furnishings

wood

Many pieces of furniture, including couches, benches, ottomans, tables, chairs, mattresses, and headboards, utilize lumber for structure and foundational support. Since we started with the home's foundation, let's also start with furniture foundations, which are often made from timber.

Hardness is arguably one of the most misunderstood things about wood in general. Technically, hardwood refers to wood harvested from a dicot tree, such as a broadleaf variety. Softwood, on the other hand, comes from a gymnosperm tree, such as a conifer. It's not always a reference to the wood's ability to withstand force, scratches, or dents. Softwoods include timber from fir, pine, and cedar, whereas hardwoods include cherry, oak, walnut, and maple, among others.

Hardwoods tend to be sturdier than softwoods and do not contain as many terpenes and aromatic substances. Pine or cedar smells are examples of terpenes being released into the air. Though not a toxic substance, terpenes are considered VOCs and can become irritating with chronic exposure, especially for those who are more easily susceptible to allergic sensitization. Softwoods also release substantially higher amounts of naturally occurring formaldehyde than hardwoods. For this reason, we recommend going with solid hardwoods for furniture or anything that will not be sealed up.

Manufacturers may advertise furniture as being made of hardwood when the inner construction includes MDF, OSB, softwoods, or particleboard. These materials are inexpensive and easy to manipulate during the manufacturing process. However, they also release harmful amounts of formaldehyde for a long time.

The most important thing to look for is that furnishings are made from solid hardwoods. This does not include engineered or veneer wood, which is often mislabeled as solid wood or has the word veneer buried in the description. Most veneered wood furniture is affixed or glued to softwood plywood or OSB, which will release noxious terpenes and formaldehyde over extended periods if not sealed properly by you, the end user. Engineered wood and wood veneers are generally not considered healthy home options. No one wants to sit on a formaldehyde factory, so be sure to ask if it's truly solid hardwood if the product description isn't crystal clear.

TYPES OF HARDWOODS

Janka hardness test

One of the best ways to measure a specific wood's ability to withstand wear and tear (and determine how hard it will be to saw, nail, or machine) is called the Janka Hardness Test, which involves pressing a half-inch steel ball cross-grain into a block of wood. The higher the score, the more resistant a wood is to dents and scrapes. For example, the industry standard for flooring is a Janka rating of 1,000 or above.

In the US, we typically record the force required in pounds-force (lbf), but elsewhere in the world you see it in kilo-Newtons (kN). Keep in mind that most woods come in multiple varieties, so many showcase a range of Janka ratings.

Let's break down the benefits and uses of the hardwoods that furniture manufacturers use most often, all of which are on the healthy list.

ALDER Janka: 590–1,690 lbf

Known for its character and beauty, alder is a dense wood with characteristic knots and veining that make each piece one of a kind—and make knotty alder sought after by carpenters worldwide. Alder is slightly darker than other hardwoods and takes stain extremely well. It is commonly used for cabinet and interior door construction and in a variety of furniture pieces.

HICKORY Janka: 1,290–2,140 lbf

Rustic farmhouses and organic settings come to mind with hickory, which is great for flooring, shelving, and furnishings such as benches and tables. Hickory has quite a bit of character and becomes a tad yellow after clear coat application.

PECAN   Janka: 1,820 lbf

MAPLE   Janka: 700–1,500 lbf

Due to its size and density, pecan is manufactured in long, wide boards and is known for being quite durable. We've seen beautiful dressers, chests, and long living room consoles crafted from pecan. Both pecan and hickory are redder in the heart of the tree and whiter toward the bark. The wood takes on an amber hue from sunlight over time, much like the look of cherrywood.

Maple wood is one of our favorites, and it's occasionally available as a grade of plywood labeled no added urea formaldehyde (NAUF). It's not only on the more economical side of the hardwood scale but also very light and neutral colored when left in its raw state. A well-sanded piece of maple will hardly off-gas at all. Some maple pieces have more character than others, which is why our carpenter hand selects maple plywood for each cabinet, mantel, and faux wood beam project. Different parts of the tree may take on a greenish hue or display gray veining that will be more visible if a clear coat is applied. Green-tinged maple makes for wonderful paint-grade cabinetry or shiplap if cut in long enough sections. Gray veining can be incorporated into the design aspects of a modern home with a rustic touch. Maple also stains beautifully if a darker color is desired.

OAK Janka: 1,060–2,680 lbf

Oak is timeless. Many a home's steps and flooring are made of oak plank, which is as strong as it is durable. Oak comes in two main varieties—white and red. White oak is lighter in color, is a bit more durable than red oak, and has natural water-resistant qualities. The look may be a bit dated for some people's tastes, but oak is here to stay. Interestingly, cork is the bark harvested from a type of oak tree found in the Mediterranean region. Sustainable cork may be harvested once every nine years, and each tree produces roughly 100 pounds of cork per stripping.

BIRCH Janka: 760–1,470 lbf

Birch, also readily available and economically priced, has a neutral, almost white appearance. It is a popular choice for cabinetry, crafts, and dovetail drawers, as it's easiest to find in plywood form. It can be stained to mimic more expensive exotic hardwoods as well. Carpenters are fond of birch, since it holds screws, nails, and adhesive well. Birch and maple are both considered paint-grade woods, but today's white, light, and bright trends make them perfect for décor pieces finished with a nontoxic matte clear coat. Keep in mind that birch may yellow a bit over time.

WALNUT Janka: 1,010 lbf

The classically dark, impressive appearance of walnut implies luxury. Its rich, deep color and characteristic fine, straight grain make it popular for desks, built-in shelving, dining tables, coffee tables, and judges paneling. Although walnut is strong and dense, it's not nearly as heavy as other hardwoods like oak. It's also among the easier hardwoods to work with, making it exceptional for carving. Walnut is on the higher end pricewise and is almost always stained to bring out the darker ebony tones.

ASH Janka: 760–2,030 lbf

Midcentury modern–style furniture is often made from ash because this wood has an attractive straight grain and hues that range from light beige to light brown or gray. It's durable, lightweight, pretty, and resistant to heavy blows. Ash is a member of the olive tree family and is gaining popularity in the furnishings world. Beds, couches, dressers, and tables crafted from ash wood can be expected to darken slightly over time. Sometimes oak and ash look alike, but ash is much more prone to rot, so it's best to keep ash furniture indoors.

CHESTNUT   Janka: 540 lbf

Chestnut describes a family of trees, the most common in North America being the American chestnut. Its shades range from pale white to medium brown, which will develop a characteristic reddish undertone with age. These trees are durable and rot resistant, but their popularity has caused a decline in available chestnut wood. Most chestnut today is wormwood, which means the tree developed blight and died, making it a perfect haven for worms and insects. The boards will have holes and worm tunnels that lend unique character.

WHITE ASPEN   Janka: 350–420 lbf

This plentiful hardwood grows in the mountainous regions of the US and comes in shades of white, blond, and light brown. It's relatively lightweight and accepts screws and nails well. In fact, its resistance to splitting and splintering is why tongue depressors, match sticks, and children's toys are often made from aspen. Both poplar and white aspen are popular picks for infrared saunas as well. The rustic character can be brought out in furniture pieces with stains and nontoxic water-based semitransparent lacquers.

BEECHWOOD   Janka: 1,300 lbf

Plentiful and uniform, beech comes in a whitish cast and is very homogeneous. It's one of the most popular woods in the furniture industry for frames and bases in upholstered furniture. Beech plywood is readily available. Since beech responds well to steam, the wood can be bent and formed into musical instruments and tools.

ELM   Janka: 830–1,540 lbf

Although elm is tough and hard, it also steam-bends easily and holds its shape well, making it great for supportive furniture parts like frames, backs, seats, and legs that need to be structurally strong. Elm also glues well and holds nails and screws without issues. Its honey brown color has some natural waviness, but stain really wakes up the character and showcases the grainy texture. Elm is used for butcher blocks, cutting boards, chairs, baskets, hockey sticks, bats, and barrels.

MANGO WOOD   Janka: 1,120 lbf

Mango wood is becoming increasingly popular for furnishings and décor. These fruit-bearing trees are all over southern Asia. When they retire from making fruit, the wood becomes suitable for harvest as hardwood. It grows quickly and doesn't require extensive processing or kiln drying. Plus, it can look strikingly similar to teak. Its dense grains can accept a very high level of polish. Whitewashed mango wood paired with a light stain is a popular style in the organic modern and Scandinavian furnishings movement. Mango wood has zero odor, is low maintenance, and is relatively inexpensive, making this versatile timber perfect for a healthy home. Mango trees are a member of the cashew family, so some people who are allergic to cashews or poison ivy may react to mango wood through skin contact with dermatitis. (Mango fruit flesh lacks the irritant found in the sap, leaves, and fruit skins.)

WHY ARE SOME ENGINEERED WOODS OK TO USE?

Although we try to avoid it, some furnishings require the use of plywood. In those cases, we look for natural hardwood plywood (with no wax treatments or oil-based finishes). Hardwood plywood is much more moisture-resistant, so it won't warp, plus you reduce the risk of off-gassing or sap oozing out later, which is often seen with softwoods. It will still use glues, though, which is why we apply our own nontoxic finishes like paint or a clear coat to seal it after purchasing. We recommend avoiding any engineered woods that use rubber, formaldehyde adhesives, and/or petro-based finishes.

TEAK   Janka: 1,000 lbf

MINDI   Janka: 1,055 lbf

Teak possesses characteristics that other woods don't have, partly because it retains its natural oils and rubber grain even after being felled and processed. It is weatherproof and develops a silvery gray patina over time. Teak's natural oils work to protect the honey brown wood from fungi and parasites. Its elegance and durability make teak pricey but worth the investment. Boats, decks, outdoor patio furniture, and shower benches are often made from teak. Indonesian Jepara furniture made by top teak manufacturing companies is exported all over the world.

Although it's not a well-known name, you'd be surprised by how commonly Mindi wood is used in furnishings and homes. It's used for interior doors, furniture bases, chests, window frames, furniture, and novelty items due to its ease of use in woodworking. More expensive types of woods are often attached as veneers over Mindi wood because of its resistance to shrinkage. As long as the veneers aren't glued on with adhesives that contain formaldehyde or petroleum-based solvents and are finished with healthy products, this is a welcome exception to our "no veneer furniture" rule.

GEEK BOX | ## reclaimed wood and going green

Remember, green does not mean healthy. As such, reclaimed lumber is a no-no for a healthy home because there is no way to know how the wood was treated in its prior life. Often, pesticides or even coal tar creosote have been applied to reclaimed wood, which are both highly toxic.

Fumigation is another process that wood and even furnishings are subjected to. This is required by law for overseas shipments to prevent the spread of bugs from one country to the next. There's not much that can be done about that, and the good news is any fumigation will likely dissipate long before the furniture arrives at your doorstep.

Does that mean there's nothing you can do to acquire more sustainable wood? Of course not! To find the most eco-friendly timber varieties, look for materials and furniture made from wood that's FSC-certified. This means the way it was grown and harvested passed the strict standards of the Forest Stewardship Council, a global nonprofit that helps promote the responsible management of the world's forests.

EUCALYPTUS   Janka: 1,125 lbf*

Eucalyptus is naturally water-resistant, very durable, and among the faster growing hardwoods. A mature eucalyptus tree can be harvested in around four years, while an oak may take several decades to mature. Eucalyptus is one of the few hardwoods that can have a distinct smell from its naturally occurring terpenes. Planks come in either solid wood or strand-woven forms. The latter means that strips of Eucalyptus are woven together and then compressed under extreme heat and pressure to fuse them into a solid, exceptionally hard plank. We usually do not recommend strand-woven eucalyptus since most resins used during manufacturing contain formaldehyde (there are exceptions). We also recommend that you apply a nontoxic clear coat if you decide to purchase furniture with solid eucalyptus components.

3,000–5,000 lbf for strand-woven planks.

CHERRY   Janka: 950–2,350 lbf

Black cherry is among the most prized furniture hardwoods in the Americas. Its smooth grain and reddish-brown hue develop a lustrous patina over time. Cherry is popular for cutting boards, kitchen cabinets, wood flooring, and wood bowls and spoons.

ACACIA   Janka: 1,700–2,220 lbf

Talk about a cool wood! Acacia is naturally smooth and has a velvety feel due to its incredible density, which beats out oak and hickory and measures closer to marble. This warm-toned light wood is excellent for countertops and dining tables, as it's naturally water-resistant and antimicrobial even without special treatments or sealers. Low maintenance is the name of the game. Even though it's water resistant, it's not waterproof like teak. Exposure to constant moisture isn't advised, as acacia will eventually swell and warp.

MAHOGANY   Janka: 800 lbf

Mahogany is expensive, mainly because it's an exotic wood with unique coloring that grows in tropical regions of the world. Most mahogany is veneer since the cost is so high. While the reddish-brown hue and gorgeous grain is second to none, it's not the best wood for a healthy home because of its tendency to absorb moisture, even after it's been kiln dried. Over time, it will shrink, warp, and crack, just like my grandma's mahogany side table that was passed down to us.

SUNGKAI   Janka: 300 lbf

Sungkai is native to Indonesia. Because it is easy to work with, it is a popular timber species for indoor furnishings. It's relatively cheap and readily available in Indonesia; however, it lacks in strength and tends to crack compared to other similar hardwoods.

RUBBERWOOD   Janka: 890–960 lbf

Natural latex rubber is harvested from rubberwood tree farms. After the useful life of the tree, these rubberwoods are cut down or burned. Since it's often viewed as a by-product, rubberwood is relatively inexpensive compared to other hardwoods. Processed rubberwood is very stable, takes stain well, and resists shrinking and cracking. It comes in a beautiful blond to medium tan color range. It's important to note that if you have a contact allergy to latex, you may want to opt for another type of hardwood.

POPLAR   Janka: 300–540 lbf

Poplar is typically a white or creamy yellow hue, though it can also be brown or streaked. It is an economical paint-grade hardwood that grows incredibly fast and is often used for cabinetry, window trim, and door casings. It dings quite easily, so it's usually not used to make furnishings. Natural poplar is an excellent choice for use in infrared saunas.

TIGERWOOD   Janka: 1,850–2,170 lbf

This visually interesting and expensive hardwood is prized for its durability and beauty. The striking bands that run through the grain resemble black tiger stripes, which contrast with the russet brown wood profile. It resists bugs, rot, and water and is extremely dense, making it excellent for cutting boards, wooden utensils, and decorative cabinetry.

OKOUME   Janka: 380 lbf

This exotic wood from Africa is exceptionally soft and light and is typically not pressure treated. Okoume plywood is generally used for marine applications since it's the easiest to bend of the marine plywoods. It's an attractive and durable choice—just make sure to source the plywood with a core made from poplar or another hardwood.

IPE   Janka: 3,510 lbf

Ipe wood is similar to teak but easier to source if you live in the US. Ipe grows in Central and South America and has become a low-maintenance alternative to expensive teak for decking and outdoor furniture. Interestingly, ipe has nearly four times the hardness of teak and is so dense that it sinks in water. In fact, ipe has a class I fire rating—the same as steel and concrete. Also known as ironwood, it's durable, insect resistant, and won't decay.

WOVEN WOODS + REEDS

We buy a lot of woven rattan furniture. Rattan is made from the rattan plant, a vine that grows in Southeast Asia. It is generally available in a natural range of beiges and browns or can be painted with colored lacquer. A common type of gray rattan is referred to as kubu, which is stained naturally through an application of mud and clay during processing.

Cane is another common woven material that comes from the rattan plant. Cane is the thinner, lighter-colored material that results from the process of stripping the rattan plant. Sometimes these woven furnishings are referred to as wicker, a broader term for furnishings that can be made from natural materials such as rattan, strand bamboo, reeds, willow, or even synthetic fibers like nylon or polyester.

Woven baskets are a cute addition to any modern farmhouse or boho-style home. Baskets are generally made from rattan, cane, wicker, or woven abaca, which is a strong brown fiber that comes from the leafstalks of the banana tree. Baskets made from wool, jute, organic cotton rope, or metal are all great choices for a healthy home.

Furniture and basket materials to avoid are those made from seagrass or water hyacinth, due to their tendencies to eventually grow mold. Even after these natural grasses are processed and dried, they can be troublesome.

Engineered bamboo or bamboo plywood is marketed as an eco-friendly option, but we don't use these materials in furniture in our healthy homes because, with few exceptions, the adhesive used in bamboo strand board contains quite a lot of formaldehyde. Any furnishings with MDF or particleboard are on the avoidance list, too. This includes MDF labeled CARB II compliant or LEED Gold, as these still allow for some formaldehyde emissions.

WOOD PROCESSING + FINISHES

DIY FURNITURE POLISH

Did you know that it's nearly impossible to find healthy furniture polish in a store? Here's a quick, healthy recipe you can follow at home.

- 2 tablespoons white beeswax
- 6 tablespoons extra virgin olive oil (EVOO)

You'll need a large pot, a mason jar, tongs, and something to mix with. First, put two inches of water in the pot and heat it to steaming (not boiling), then put the beeswax in the jar and use tongs to set it gently in the hot water. Leave the lid off the jar and make sure the water does not mix with the beeswax. Let the beeswax melt completely, then pour in the olive oil and stir well. That's it, you're done! Just leave it to cool and firm up, and then it's ready to use.

The number of wood finishes used in the furniture industry is considerable. The best way to order your wood or wicker furniture is with no finish, aka "natural." This means no paint, no stain, no lacquer, no clear coat, no wax—just virgin wood. This way, you can create a one-of-a-kind piece by adding your own nontoxic stains, paints, and finishes. We also recognize that some people have neither the time nor the interest to fool around with applying their own finishes. You can always hire a professional painter to finish out your furniture. Leaving solid hardwood furniture au naturel (meaning unfinished) is also perfectly acceptable and looks fantastic, too.

We recommend avoiding oil-based finishes and waxes unless you're sure they are 100 percent plant-based and/or made from beeswax, with no paraffin or petroleum distillates. When conditioning wood pieces, stay away from silicone-based products, as they're not ideal for the longevity of the wood. Drying oils are your best bet, as they penetrate the wood and create a hardened polymer network. Soy, walnut, and refined hemp oil are also excellent options and can be used on butcher block countertops, wooden toys, and food-safe surfaces. We don't use linseed or tung oil because these off-gas aldehydes and have a persistent odor.

In lieu of harsh chemical stains and paints, we sometimes use natural coloring alternatives, such as tea, vinegar, and steel wool concoctions to add color and stain to wood. Some manufacturers of solid hardwood furnishings also incorporate traditional types of stains such as clays and vegetable dyes, which offer beautiful, ethical furnishings. Fair trade, eco-friendly, and

sustainable are good descriptors to look for when sourcing pieces with nontoxic finishes. However, keep in mind it's just a place to start, and you still need to do additional homework for each piece.

It's best to avoid sourcing any furniture that has varnish, shellac, or a color-painted lacquer coating, even if it's labeled as water-based. Water-based does not mean healthy—it just means some of the ingredients used to make the paints were water-soluble. The Finishes chapter includes very helpful information on what to look for when it comes to paints, stains, and finishes. Avoid epoxy coatings as well. Find coatings that are rated zero for HAPs.

Clear nitrocellulose coatings (NCs) are commonly found on wood, wicker, and woven grass furnishings. This automotive-type of clear polishing lacquer cures very hard and rapidly; it does utilize toxic solvents that we do not recommend, but the NC curing process essentially bakes the chemicals into a hardened shell. These high-gloss coatings render a glass-like finish on wood, and we've found most to be surprisingly tolerable. Any off-gassing is not enough to warrant concern for a few pieces of furniture (in other words, don't fill your home with NC-coated materials). That said, there are a number of NC coating manufacturers with varying processes and ingredients, so we recommend trusting your nose and using your best judgment.

kiln drying

One of wood's distinctive characteristics is called hygroscopicity, which refers to a material's ability to absorb or desorb moisture during changes in humidity. All types of wood do this until they reach equilibrium, which describes a balance between the wood's moisture content and that of the surrounding air. This can change a piece of wood's dimensions.

As such, kiln drying is essential for many species of lumber, and the process involves both artistry and science. Wood is dried in a giant chamber (picture a huge oven) where air circulation, relative humidity, and temperature can be controlled so that the moisture content of the wood can be reduced to a target point without having any drying defects. Because the dimensions of wood products change with fluctuations in relative humidity, kiln drying is essential for being able to machine, glue, and finish the timber later.

Although it is a bit pricier than wood that hasn't undergone this process, the result is a lighter-weight, sturdier piece of wood that resists biological deterioration from fungi and insects.

JOINERY

All manufactured furniture must be assembled, either in the factory or by the end user—you. A mechanically fastened joining process is superior in quality and preferable over the use of adhesives.

Dovetail drawers are a prime example of an expert carpenter's work, where the lumber is manipulated to fit together like a jigsaw puzzle. Other examples of mechanical fastening include nails, screws, and clips. Adhesives as joining materials are often a sign of cheap or poor-quality construction. There are certain applications where nontoxic wood glue is mandatory (such as in cabinet making), but most high-quality construction uses hidden fasteners and tiny nails.

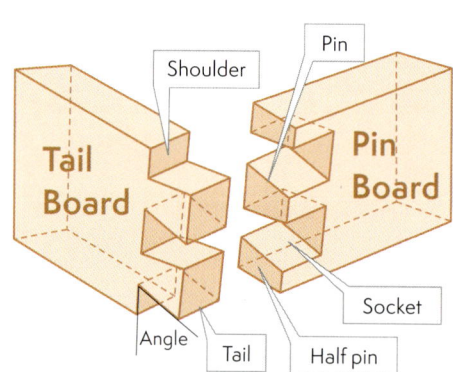

key points

▸ We recommend looking for furnishings made of solid hardwood; use the hardwoods list in this chapter for easy reference.

▸ Many types of hardwood are available, so do your research. (Maple is one of our favorites.)

▸ In addition to ensuring woods are safe for a healthy home, check the Janka hardness scale to see how resistant it is to dings and dents. Anything above 1,000 is usually good for wood furniture that needs to be durable.

▸ Woven furniture made from rattan, cane, wicker, or woven abaca is usually a great choice, but avoid moisture-loving seagrass and water hyacinth.

▸ There are lots of wood finishes to choose from. We recommend ordering solid hardwood furniture "natural" and then applying your own nontoxic stain, paint, or sealer.

▸ When it comes to joinery, we prefer mechanical fastening (e.g., nails) over adhesives.

Ready to read more about furnishing materials? Head to Chapter 17 for all things metal and stone.

metal
+ stone

Metals are known for their structural strength, durability, and load-bearing capacity and are used in furniture for support. Metallic components are also used for decoration to lend character and glamour. In fact, metal can often define the style and date of a piece. Think weathered iron for a rustic barn house or shiny stainless steel for a modern industrial look.

Common Alloys

STEEL
A combination of iron (metal) and carbon (non-metal)

BRONZE
A combination of copper (metal) and tin (metal)

BRASS
A mixture of copper (metal) and zinc (metal)

Our recommendations concerning metallic furnishings are relatively straightforward, since metal is an inert material in most respects. (Metallic alloys such as steel, bronze, and brass are mixed metals or compounds.) Be aware of whether you or someone in your family has contact allergies to specific metals when purchasing seating, tables, handle pulls, or anything that might be touched. For example, if someone in your family has a sensitivity to copper, then bronze and brass are likely not a good idea. And of course, avoid radioactive metal alloys. We recently came across a company that claimed their lamps made with radioactive elements would neutralize EMFs within a home. As we discussed earlier in the book, many types of radiation are toxic and can cause cancer. We couldn't believe these lamps could be sold legally!

Popular metal surfaces include stainless steel and copper, the latter of which is pricey but beautiful and naturally antibacterial. Most metal surfaces like stainless steel are available in fingerprint-resistant finishes and are easy to clean with a mixture of water and a tablespoon of phosphate-free/nontoxic liquid detergent concentrate. (This type of soap is the same stuff used for construction cleanup and degreasing metal HVAC ducts that we referenced in the Air chapter.) Copper surfaces require only a wipe-down with a barely damp cloth. We don't use store-bought stainless-steel cleaning products.

Some furnishings are labeled as containing lead, which may be acceptable in certain situations. If you're buying a lamp or light fixture that contains some lead but is going to be out of reach, it's likely not going to be an issue. However, if a whitewash on a woven vase that you may handle from time to time contains lead, you should probably skip the purchase. Why? Because lead is a highly toxic heavy metal that is surprisingly prevalent in home goods. Lead mimics calcium inside the body and will be recycled and stored in a person's or pet's bones for decades once absorbed.

WHY IS LEAD SO PREVALENT?

Lead has a low melting point and is very pliable (bendy) compared to other metals. This advantage made it a preferred ingredient in manufactured metal alloys that were designed to be used as flashing (for example, roof waterproofing), coatings (such as paint), and molded products (for example, plumbing pipes). Chemical engineers also discovered that lead helped prevent knocking in combustion engines, so all car and jet fuel contained the additive as well.

But lead is known to damage the brain. Even scarier is that children are particularly vulnerable to the horrific effects of ingested or absorbed lead and are at the greatest risk for irreversible brain damage. Although these effects have been known for centuries (they were blamed in part for the fall of the Roman Empire), the application of lead in consumer products is still legal in many parts of the world.

Lead paint was federally outlawed in the US in 1978, and Congress later banned the use of leaded pipes in 1986—but allowed those already in the ground to remain. Over three decades later, an estimated fifteen to twenty-two million Americans still cook with and drink tap water that enters their homes through lead pipes, known as service lines.[1]

Leaded fuel was banned for use in automobiles in the US in 1996, but aviation fuel still contains lead, and plane engine exhaust disperses copious amounts of lead dust into the air we breathe 24/7. What many people don't know is that home products like crystal glassware, glazed dishes, jewelry, and children's toys may also contain lead. Heck, leaded solder is still available and as legal as ever—another excellent reason to use PEX pipes instead of metal ones for water delivery at the point of use.

We look for manufacturer claims that confirm the absence of lead when purchasing pots and pans, glassware, dishes, ceramics, and painted toys. Pewter is an alloy that we do not allow in any of our healthy homes since it often contains lead. Be mindful of bargain-priced silver, as the potential for lead is high.

Do your homework when purchasing products with baked-on enamels, as some resins used in this process can flake off and/or release chemical fumes into the home, especially when heated (in the oven or direct sunlight).

Another similar-looking metal finish is powder coating, which has superior durability and is an excellent healthy choice. Powder coating is applied through a special electrical process and will adhere to metal and aluminum. Powder-coated shelving units are ideal for the wear and tear that comes with storing sporting equipment and tools in the garage. We suggest using metal peg boards for tool storage, since traditional peg boards are made of MDF and particleboard.

Remember to wipe down metal chairs, tables, fixtures, and accent pieces before bringing them inside. Sometimes petroleum grease is still present on metal pieces after being made in a machine shop. A 2:1 solution of clean water and clear unscented ammonia is a great degreaser. Spot test this first, though, as some metals like aluminum are not suitable for ammonia applications. A spray bottle of clean water mixed with one tablespoon of phosphate-free/nontoxic industrial grade detergent concentrate is also great at breaking up and removing grease.

WHAT ABOUT TOYS?

Toys for kids and pets should be as durable as they are nontoxic. Both groups put everything in their mouths, and some options can be harmful to your littlest family members.

Look for wooden toys and puzzles made from balsa or solid wood rather than pressed wood, the kind with lots of lines in the wood grain. The glues used in pressed woods contain phenol formaldehyde. If you get metal toys, make sure they are lead-free. Paints should of course be labeled nontoxic, and check to make sure small pieces can't be taken off and choked on.

Silicone toys with no holes are great bath time entertainers. The rubber duckies we grew up with, however, are notorious for growing mold inside the cavity. Although so many toys are made from pliable PVC or BPA plastic, opt for toys that are built with hard plastics, like polypropylene, instead.

EMF POTENTIAL

We've heard quite a bit of concern expressed about EMF amplification and metal furniture. People ask whether it's OK to sleep on a metal bed frame, by metal side tables, and/or on a mattress that sits atop a metal box spring. These are legitimate concerns if you're EMF-sensitive, but there's more to it, as this is only a problem if you already have EMF issues in this area of your home.

We build healthy homes according to our own EMF best practices, which include routing bundles of wiring behind walls and through subfloors on multi-story residences away from bedroom walls where people sleep. Furthermore, our homes are designed to run using wired components only, eliminating Wi-Fi, Bluetooth, or other constantly broadcasting wireless signals. So, inside a true healthy home, concern over metal in the bedroom should be minimal.

An interesting side note to this conversation pertains to EMFs caused by the Earth and outer space. Sounds sci-fi and silly, right? However, it's as real as the laws of physics. But most people don't know about this interesting environmental phenomenon happening around us daily. If you have certain types of EMF sensitivities and you live in an area that produces it, geopathic stress may become problematic for you.

What exactly is geopathic stress? Some people believe that geopathic forces are natural vibrations from the Earth. It's a natural phenomenon our bodies are inherently equipped to handle. Most people don't feel any physical effects of these energies, but for EMF-sensitive people, surges (called geomagnetic storms) can cause physical stress that may undermine the body's subtle energy systems. This is especially true in individuals who have an altered or sensitized limbic system response and have become sensitive to certain electromagnetic frequencies. As we talked about before, these hypersensitive responses are commonly linked to toxic mold and chemical exposures. We are electrical beings (as discussed in the Energy chapter), and our hearts, organs, muscles, brains, and nervous systems run using a grid of thousands of energy pulses generated every second. So, it makes sense that energy from the Earth can interfere with the human body.

In terms of avoidance, excessive geopathic stress is thought to happen in areas along the Earth's surface where natural EMFs are magnified—that is, in regions where there is movement of tectonic plates, such as along fault lines, and in regions rich in underground crude oil reserves. Spikes in electromagnetic energy also happen in the hours or days preceding earthquakes, which is how early warning systems work. Interestingly, land over oil reserves or fault lines is often barren. The lack of trees and abundant plant growth is perhaps a subtle sign that something is making it hard for life to thrive. The basic takeaway here is to avoid spending time in areas prone to geopathic stress. Although it's hard to chart these areas on a world map, the rule of thumb is simple: don't build your home near fault lines or on top of underground oil reserves.

Space weather and solar storms are another natural phenomenon that most of us don't pay any mind. Our bodies are equipped to handle this type of radiation as well. An example of this is occasional GPS signal loss as a result of a solar flare. Some EMF-sensitive people, however, may feel the stress from solar storms, so this would be another reason to avoid metal furniture or beds with metal components. A metal roof may block this kind of energy from entering the home as it travels through the Earth's atmosphere. (More info is available in Chapter 8 about metal roofing installations.) If you're still concerned, some companies online sell metallic fabric canopies that can be draped over the bed to help block radiation.

STONE AND OTHER SOLID SURFACES

Solid surfaces are a functional component for tables, chairs, shelving, dressers, and more. These all have solid surface areas where you can place decorative items, lamps, or whatever you need within arm's reach.

Clear acrylic, solid hardwood, glass, metal, and stone surfaces are all excellent options. Hard plastics with no PVC are also safe for the home and are good options for children's furniture. Manufactured quartz (the same type that is used in countertop installations) contains acrylic-based resins and natural minerals, which are durable, healthy, and beautiful. Adding a dining table or desk with a top made of marble, travertine, slate, or quartzite is certain to make a stylish statement. Ask the manufacturer if you can order the piece without stone sealer, since most manufacturer-applied sealers are PFAS-based.

Glass is also a great option, as it's easy to clean and not as pricey as stone. Avoid glass with spot-resistant coatings or tinted surface treatments. Unfinished concrete is another nice option if you apply a nontoxic sealer to the surface to help prevent staining. Penetrating concrete sealers are beneficial, but they won't work as well for stain resistance. Walnut oil is a good drying oil for concrete countertop maintenance. Remember to spot test first.

Some stone surfaces are porous and will require a maintenance application of a sealer. These include marble, travertine, soapstone, limestone, sandstone, onyx, and slate. Most stone sealers on the market are made of fluoropolymers, which are on the naughty list. You can use either a nontoxic clear sealer for lighter-colored rock or several coats of a finishing oil on most of the darker-colored surfaces that do not have a gloss finish. Walnut oil is a great option that can be found in the cooking oils section at the supermarket, which means it's safe for food prep areas. However, those with nut allergies should avoid using it, along with shea butter and other nut oils. Other pantry staples, such as olive or canola oil, won't work in this application, as they can go rancid. Placing a droplet of water on the surface is an easy porosity test. If the water is absorbed in less than a minute, then it's time to apply a seal coat.

It's important to note that real natural stone surfaces can degrade when vinegar or hydrogen peroxide is used for cleaning. The best cleaning product for stone is one with hypochlorous acid as the active ingredient mixed with water. It is effective for cleaning and disinfecting if you spray and leave it on the surface. We use it on our kids' toys and play areas and even as hand sanitizer. It has hardly any scent and breaks down into harmless saline.

We stay away from furnishings with granite surfaces for indoor use, as some granite has the potential to be radioactive. Avoid epoxy surfaces and resins or composites not made from acrylic. Some laminate surfaces are OK if they are made from polyethylene and are completely encapsulated on all sides, surfaces, and seams. However, keep in mind that bargain laminated composite materials can crack or peel over time and may cause chemical fogging via chlorinated plastic fumes. Surfaces with rubber (synthetic latex) are not recommended due to off-gassing. Faux marble products, sometimes referred to as cultured marble, are also cheap types of laminate that we avoid.

key points	

▸ Metal can be a great choice for furnishings, as it's mostly inert.

▸ However, be wary of metal allergies, and know what a metallic alloy (mixed metals) contains.

▸ Always avoid anything labeled as containing lead if it's within reach or can create dust, as lead is a dangerous heavy metal.

▸ If you've set up your home to be otherwise healthy, a few pieces of metal furniture shouldn't cause EMF concerns.

▸ Alternatives to metal include clear acrylic, solid hardwood, glass, and stone, but make sure you know how to treat your choice with nontoxic sealers and cleansers.

In our final chapter on sourcing furnishings, learn all you need to know about fabrics, upholstery, foam, and everything in between.

CHAPTER 18

fabrics, upholstery, foam + more

There's a dizzying array of fabrics available for everyday household items like upholstered sofas, decorative pillows, and window treatments. In fact, many people base their furniture-buying decisions on the color, texture, and look of an upholstered piece.

DIY LEATHER CLEANER

Combine three parts rendered tallow with a little white vinegar to add some cleaning power to the mixture. Whisk and apply with a clean, dry cloth. You can find 100 percent pure, clear beef tallow online marketed as a fragrance-free skin moisturizer. Always spot test first in an inconspicuous location.

One of our favorites for its beauty and durability is leather. Most leather is made from cowhides, but some is riddled with toxic dyes, solvents, and stain treatments. Custom-ordering leather furniture is the way to go and lets you select your color from options that are aniline dyed with a water-based formula. Nubuck leather is super soft to the touch and is by far the least processed type, as it doesn't have a layer of protection on top. It's better to go with darker colors so stains and spills aren't noticeable (just make sure to remove spills promptly with a damp rag). Request that your custom leather piece not have any stain treatments applied. While all leather furniture pieces will have that new leather smell, in the types we recommend it will be faint and will dissipate after several months. Remember to condition leather periodically, which is easy to do with a simple homemade leather cleaner.

 GEEK BOX | the demand for viscose

Demand for viscose is growing by approximately 5 percent each year, and roughly ten factories were producing 70 percent of the world's viscose as of 2017. So why are powerful fashion brands pushing for less fragmented, more sustainable viscose manufacturing processes? A few plants are using dirty practices that pollute the environment and jeopardize the safety of workers, while other factories are producing clean viscose using responsible manufacturing techniques that involve a closed-loop process that avoids environmental contamination and is mindful of worker safety. Dubbed "art silk" (short for artificial silk), viscose is actually not a synthetic material but is cellulose-based and takes on the unique softness and texture of silk for a fraction of the price.

OEKO-TEX

This strange-sounding designation means quite a bit in healthy furnishing circles. If you see OEKO-TEX on a label, that textile product has undergone extensive certification to ensure not only sustainable manufacturing but also limited or zero harmful chemical content.

This is the work of seventeen independent research and test institutes located across Europe and Japan, whose goal is to create a safer, healthier, and more sustainable world for everyone. What's more, they offer an online label checker and buying guide so you can be sure the item you're thinking about purchasing is legit.

Natural fibers are our favorites for fabric upholstery. Look for organic cotton or OEKO-TEX certifications for cotton and linen furniture products. Viscose is gaining popularity and is a personal favorite. Rayon is the same thing as viscose; they are both cellulose-based materials that are typically made from bamboo or beechwood. Viscose has excellent temperature-regulating properties, is breathable, holds color well, and is super soft. The material does tend to absorb stains if not removed promptly, and it's recently received some bad press regarding the chemical processes involved during manufacturing (at a few plants, not all).

Lyocell from eucalyptus and modal from beechwood are other fabulous alternative fabrics with similar properties to rayon. Hemp and Belgian flax linen are nontoxic options but aren't as soft as the aforementioned fabrics. If you're looking into using genuine silk, make sure to ask if it's been treated with pesticides. Silk comes from the silkworm, so pesticides are often used in the process.

Wool and alpaca are other great fabric options, and wool is naturally fire retardant. Most of the natural fabrics (except for cotton) are not treated with fire retardants, but you should always ask. Stay away from materials that have stain-resistant finishes, as these are often made from the same toxic nonstick material we warned against earlier in the book, commonly referred to as PTFE (polytetrafluoroethylene). PTFE is readily absorbed through skin contact and is linked to cancer. Waterproof and wrinkle-resistant treatments are often made from PTFE as well. Your best bet for outdoor patio furniture cushions is to store them inside or in a weather-tight box when not in use.

Conventional cotton (100 percent cotton) that isn't labeled organic or OEKO-TEX certified likely contains pesticide residues, so we advise against purchasing standard cotton fabrics. Other downsides to all cotton (organic and conventional) include its inability to breathe compared to other natural fibers, and it can be contaminated with mold, especially the Alternaria and Aspergillus species, which are sources of mycotoxins and allergens for some.[1] In fact, it's not uncommon for people with a history of living in a mold-infested home to become allergic to wearing cotton and sleeping on cotton sheets.

Nylon and polyester aren't necessarily on the avoidance list, but it's important to note that they are not ideal for a healthy home. They're synthetic materials that don't breathe well. Natural materials are much easier on sensitive skin and offer superior comfort, beauty, and thermal performance.

Vinyl that is labeled as polyvinyl acetate (PVA) is economical, waterproof, and great for upholstery applications. Avoid PVC coatings or leather labeled PU or faux leather, since PU stands for polyurethane, a known off-gasser.

If an upholstered piece is labeled dry-clean only, check whether it can be steam cleaned. Nothing should ever be dry cleaned, due to hydrocarbon solvents and carbon tetrachloride residues that are left behind following the process. These are readily absorbed into the skin and may cause cancer or damage the central nervous system, liver, and kidneys.

what's the deal with dry-cleaning?

Most dry cleaning isn't dry or clean. The process uses a liquid solvent that typically involves a chemical called perchloroethylene, aka tetrachloroethylene or "perc."

Although highly effective at removing stains and scuffs from clothing, perc is readily absorbed via skin contact and has been classified by the EPA and the International Agency for Research on Cancer (IARC) as a likely human carcinogen, meaning that prolonged exposure increases the risk of cancer. And yet it's still widely used across the industry. The "why" is a matter that's up for debate, but we know one thing: we're not taking our clothing or upholstery in for dry-cleaning.

FOAM + FILL

Shoppers often give little thought to things like the stuffing used in couch cushions, decorative pillows, and headboards. After all, they're not visible, and it takes mental effort to break the out-of-sight, out-of-mind mentality that many of us have.

Truth be told, most of the foam in mattresses and upholstered products is made from standard polyurethane foam, which can release VOCs. Furthermore, conventional furniture foam is often treated with chemical flame retardants and contains heavy metal impurities.

Thankfully, CertiPUR-US is a nonprofit certification program that certifies foams (including polyurethane) to be free of harmful chemicals. Look for upholstered foam items that have the CertiPUR-US label on the tag. Some newer foams on the market are derived from corn or soy; however, most of these manufactured products include a percentage of polyurethane foam to add stability. Look for the CertiPUR-US label on these as well.

Baby play areas and gyms are also areas where interlocking foam flooring is a good idea. However, we have been thoroughly disappointed in the quality of the foam flooring available. Most ethylene vinyl acetate (EVA) flooring kits off-gas strongly, even after months of baking in the sun and being washed. Alternative click products for gyms include a linoleum-type flooring, which has been around for a while and is made of natural materials. However, we caution against the use of this product because the slurry used to make the cementous mixture contains pine rosin and linseed oil, which have a rather strong and unpleasant odor that can persist and turn problematic with chronic exposure. Cork is still the best and only option we are aware of for resilient and water-resistant gym floors and baby play areas, where the goal is to protect the subflooring from damage and/or cushion the flooring during exercise or play activities.

Most pillow and comforter stuffing is polyester fiberfill. It may be marketed as hypoallergenic and good for the planet (if it's derived from recycled plastics), but both are wrong. Recycled plastic can be recycled clothing detergent bottles, which means your pillow will smell like a laundromat. As for green, well, the poly industry uses tons of oil each year in manufacturing polyester. By-products from the manufacturing include formaldehyde and benzene, which are released into the atmosphere. While it's not detrimental to indoor air quality to have items in your home that are made with polyester fiberfill, it's still a petroleum product that can release tiny airborne lint particles that we shouldn't be breathing. When given the option, it's best to stick with nontoxic animal- and plant-based fillers.

CertiPUR-US

This certification program is administered by a nonprofit using independent, accredited testing laboratories to ensure foam is made without:

· PBDEs, TDCPP, or TCEP flame retardants

· Mercury, lead, and other heavy metals

· Formaldehyde

· Phthalates

Even better, they ensure low VOC emissions for every piece of certified foam (less than 0.5 part per million). Despite our zero-VOC rule, sometimes it's unavoidable, and that's pretty darn good.

Feather fill is one of the best insulators around and offers an especially fluffy body. Feather down is a premium product and may cost a bit more, but its look and feel can't be rivaled. A new standard has emerged for ethical sourcing of feathers called the Responsible Down Standard (RDS), which prohibits force-feeding and the removal of feathers from live birds and audits each stage in a retailer's supply chain to ensure that down and feathers come as a by-product of food processing from healthy animals. We're all about ethical treatment of furry and feathered friends. Those with feather allergies should of course avoid feather stuffing.

ALTERNATIVE STUFFING

Impressive alternative stuffings are becoming more readily available, such as wool, organic cotton, and PLA (derived from non-GMO sugarcane). Kapok is a widely available vegetable-based fiber that's made from the seed pods of Kapok trees. Because it's not farmed but rather grown in the wild, Kapok fiber requires no pesticides to grow. It's also hypoallergenic and mold resistant.

Seeds such as flax and millet are used in head and neck pillows for their resiliency and ability to be shaped. Of the two, flax has an advantage since it's scent-free. Natural latex filling is harvested from the milk of the rubber tree and is used in foam and fill applications because of its springiness. Natural latex can also have a faint odor that some people do not care for but is tolerated well by most. All these fills come from plant-based, rapidly renewing resources, which is also good for the planet.

Some newer alternative stuffing materials found in duvet blankets and sleeping pillows include bamboo floss and lyocell (aka Tencel) floss, which are ultra-soft and offer the unique property of thermal regulation. In other words, your body will stay cooler under the covers when it's warm and warmer during the cold winter months. Some DIY ideas involve repurposing dryer lint or cutting up old clothes as fill. You can always buy a pillow and restuff it with your choice of filler later.

Hulls from buckwheat and walnuts are also fillers found in specialty head and neck pillows, but they both have a grain-type smell and could potentially be contaminated with mold. Also, stay away from anything that resembles those white foam disposable cups. Polystyrene beads are often used in bean bags and poufs as a cheap filler, but they release styrene, a potent VOC that messes with the central nervous system and is attributed to headaches, fatigue, and difficulty concentrating. Molded polystyrene is also found in children's car seats. If there's a high quality, safety-tested seat with alternative materials available, opt for that. That said, we don't recommend buying an untested seat just because the label seems healthier. Safety first.

Let's not forget that our furry friends also need a comfy place to snooze. The same considerations for upholstery materials, fillings, and foams apply. Cat towers made from jute or natural rope materials are a healthy alternative to carpet. As tempting as it was for us to get one of those cardboard scratching boards for our cats, it was a hard pass because cardboard is loaded with formaldehyde and should stay out of the house.

BEDDING + MATTRESSES

Although we've touched on mattresses and bed framing materials elsewhere in the book, there are still some details worth highlighting. Mainstream mattresses are treated with arsenic and other chemical flame retardants. They also have toxic foams and could be made with softwood (for example, pine boards) for structural support. We spend a third of our lives sleeping, and a restful night is imperative for optimal functioning. So, an investment in a nontoxic mattress is a must.

Look for brands that are transparent about their products and processes. Organic cotton, natural latex, and wool are fabulous mattress materials. If foams are present, look for the CertiPUR-US label. Metal or solid hardwood frames are equally sturdy and won't affect indoor air quality. Be sure your wood is finished with nontoxic, formaldehyde-free paints, stains, or clear coats that are zero VOC and zero HAP. Don't forget that a mattress needs to breathe, so a slatted bed frame that allows air flow is preferred.

The healthy materials listed in the upholstery section are the same types of materials you should seek out for sheets, blankets, and pillowcases. A mattress protector will extend the life of your mattress and should not have an integrated layer of plastic. Water-resistant organic cotton and bamboo protectors are available without a PVC film layer.

TOWELS + BATH MATS

Bathrooms are a functional space and should be designed and decorated with water resistant furnishings. Bath mats made from solid virgin teak (not the look-alike engineered bamboo) are excellent for absorbing water as you step out of the shower. They clean up easily when mopping the floor and are naturally antibacterial.

Carpeted bath mats, on the other hand, can harbor dirt, dust, and dander and really aren't designed to get wet on a daily basis. Make sure any mat you purchase doesn't have synthetic rubber for the anti-slip layer on the back.

Bamboo rayon towels reign supreme in the towel lineup; they are way more absorbent than cotton and don't harbor odors from sweat. Plus, they're oh-so-soft and outlast cotton.

WINDOW TREATMENTS

Privacy is important, especially in bedrooms and bathrooms. Finding healthy window coverings is also essential to a nontoxic living space because windows are regularly subjected to heat and UV light. The best window coverings are made from natural woven bamboo reeds or woven grass blends. These types of blinds are versatile in design and can be a great addition to any home, whether the style is boho chic, clean and modern, traditional, or farmhouse. Make sure the natural blind materials don't have any toxic dyes applied to them, and skip the expensive motorized versions, which are a source of EMFs.

Metal louvers may be old school, but they come in various colors. Curtains made from Belgian flax linen or organic cotton are readily available and should not have any added coatings. The downside to fabric window treatments is that they can get dusty and should be laundered periodically. (Pro tip: look for shower curtains that are made from flax linen or organic cotton as well.)

Window covering materials to avoid include wood composite, which are often found in plantation shutters. These are made of either plastic or an MDF-type material that releases VOCs, especially during the hot summer months. Window treatments that are designed to hold a specific shape (for example, pleated shades) are likely treated with chemical coatings and made of unhealthy vinyl, both of which will pollute the living environment.

RUGS

Rugs are an amazing way to frame a space, adding color to a room and a soft walking surface. They're fabulous for padding a toddler's falls, and they also help dampen noise. When it comes to materials, naturally derived weaves are the name of the game. Area rugs made of wool or viscose are the easiest to source and come in thousands of patterns, sizes, thicknesses, and styles.

Rugs should be vacuumed regularly (if the manufacturer recommends) to remove dust and shedding. Most manufacturers also suggest taking rugs outside occasionally to beat dirt and dust off with a broom.

We've found that wool and viscose rugs generally do not contain fire retardants or stain coatings. Don't assume this is the case, though, as laws concerning fire retardants are evolving. It's best to find rugs made of cotton certified by OEKO-TEX to be sure they're free of fire retardants and chemical additives. Silk rugs are also a great option if you can verify that no pesticides were used in the production of the silk. This one is important, as rugs are often heirloom pieces. Pesticide exposure can have dire consequences for your family's health.

Natural fibers like jute and sisal can be woven into rugs and are known for being able to take a beating. They are great for the modern organic look, and they're relatively inexpensive compared to other rug types. Some manufacturers recycle old leather pieces from the manufacturing process and weave them into floor coverings. These are cool and unique—just be sure the leather isn't one of the types that's processed with toxic solvents (see the leather section earlier in this chapter). Real animal hides are soft, textured, and safe to use, depending on the backing. Polypropylene rugs are preferred for indoor/outdoor use and can easily be washed off with a hose and line dried.

Rug backings should be fabric, cloth, or natural latex. Avoid rubber backings and anti-slip rug mats unless they use natural latex. Synthetic rubber has an unpleasant odor like car tires. Polyurethane anti-fatigue mats are not advised inside the house, and neither are large rubber welcome mats (although they are fine to place outside).

INDOOR FOLIAGE

If you're a fan of indoor plants, skip the soil and go for hydroponic planters. Bringing soil indoors can introduce mold spores that are able to multiply once they land on a surface with a food source and moisture. HVAC ducting is prone to contamination from indoor plants.

Pesticide-free silk plants are another excellent, maintenance-free option. Pesticides have a pungent peppery smell. One client of ours recently bought dozens of silk floral arrangements online—and the "plants" are still sitting outside in hopes that their strong scent will dissipate. Clean silk can be hit or miss in terms of quality, so it's worth being able to see and touch the faux plants before purchasing. Avoid fake florals made from PVC (they smell like beach balls). Dried flowers preserved with plant-based preservatives or glycerin are another good option.

We love the natural look of twisted willow branches in a large vase. Dried bamboo is also a great addition to this kind of arrangement. Ethically sourced feathers in a vase are also an interesting alternative to faux florals. Materials to avoid include woods that were not properly dried and have visible mold or lichen growth. Also, some grasses such as dried cattails or bunny tails may irritate seasonal allergies.

key points

▶ There are lots of fabrics to choose from. Natural, untreated leather is one of our favorites, as are cellulose-based natural fabrics.

▶ When sourcing cotton fabrics, look out for the OEKO-TEX certification on labels to ensure you're getting a healthy, high-quality product. If you're sensitive to mold, keep in mind that cotton can be contaminated with mycotoxins.

▶ When it comes to foam, look for CertiPUR-US on the label. Trust us.

▶ Avoid fiberfill stuffing. Go with wool, organic cotton, bamboo, or PLA instead.

▶ Don't buy a mattress treated with chemical flame retardants. Instead, look for organic cotton, natural latex, or wool.

▶ In the bathroom, we recommend teak for bath mats and bamboo rayon for towels.

▶ For area rugs, go with wool or viscose. And don't get anything with rubber backing; choose fabric, cloth, or natural latex.

▶ Some fake indoor plants are better than real ones, as the latter can introduce mold into the home.

Next up, get ready for Part 6, your guide for what to do with an existing home.

We've talked a lot about what to do to ensure you're building a new home in the healthiest way possible—that's the whole point of this book! But what if you're not currently interested in building a new home? Or you're looking to buy a home that's already built?

We always encourage you to build new for maximum control over what goes into your healthy home, but if that's simply not an option, this part can serve as a guide to ensuring your living space is as healthy as it can be.

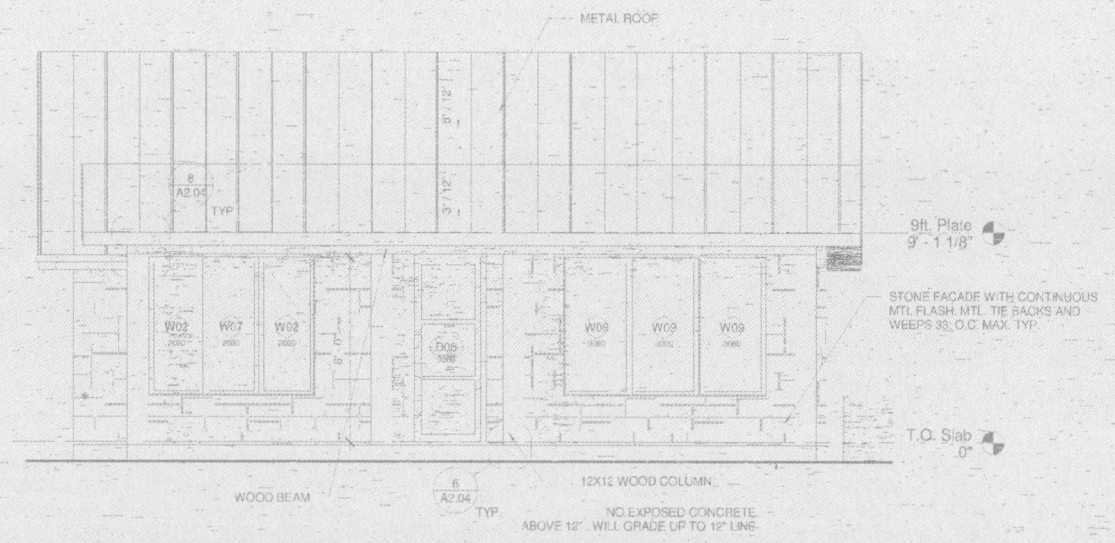

METAL ROOF

PART 6:

already home?

2 Back Perspective

make an existing home healthier— starting today

So, how do you start making your current home healthier? Let's go room by room and piece by piece through the house, starting in everyone's favorite space: the kitchen.

KITCHENS

The word "healthy" should describe not only the foods we eat but also the tools we cook with, the materials used to build out the kitchen, the ergonomics and flow of the space, and more. You can have great intentions, but if you want a truly healthy kitchen, you need to pay attention to the bigger picture.

First, familiarize yourself with the "kitchen work triangle," a concept developed to ensure orderly layouts in compact kitchens. It describes how the major work areas in a kitchen, which include the refrigerator, the stove/oven (often called the range), and the sink, should always be four to nine feet apart in something of an equilateral triangle, where the sum of the three sides is between thirteen and twenty-six feet. The idea is that if the distance is too small, the kitchen will feel cramped, but if it's too large, cooking can become a hassle. Having an ideal amount of space between the main work areas streamlines traffic and simplifies food preparation.

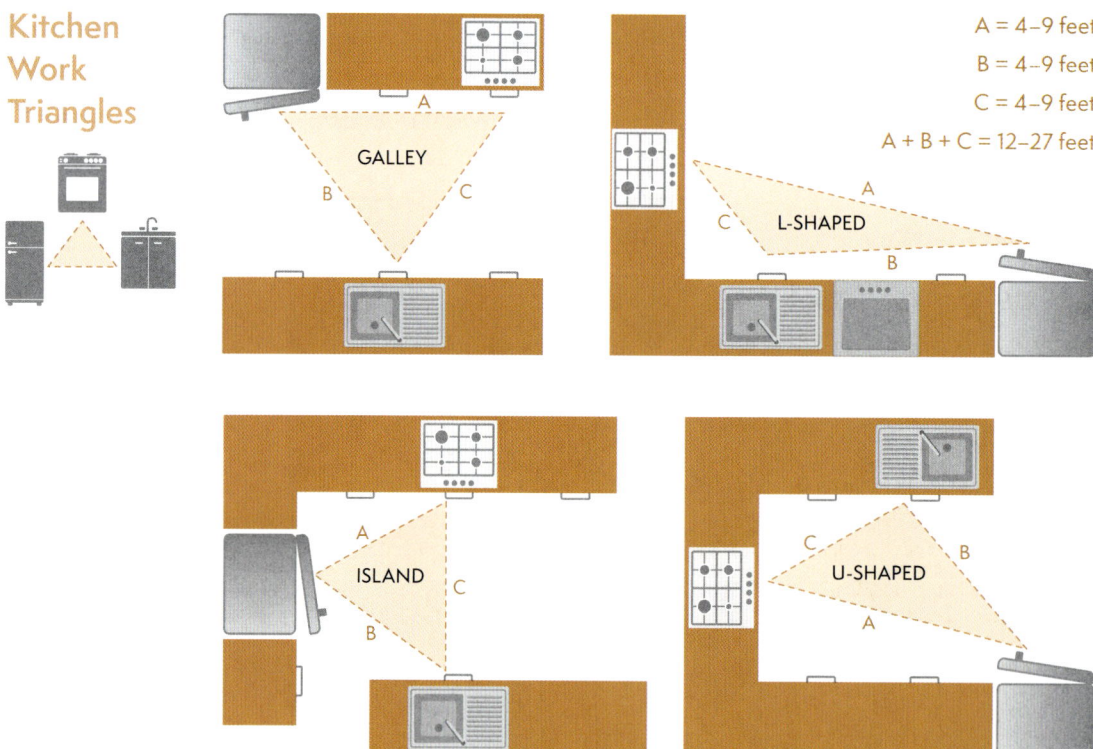

Kitchen Work Triangles

GALLEY

L-SHAPED

ISLAND

U-SHAPED

A = 4–9 feet
B = 4–9 feet
C = 4–9 feet
A + B + C = 12–27 feet

ENVIRONMENT	FOOT-CANDLES	LUX
Living room	10–20	108–215
Bedrooms	20–50	215–538
Kitchen – general	20–50	215–538
Kitchen – preparing/cooking	50–100	538–1076
Dining room	10–20	108–215

In terms of lighting, the Illuminating Engineering Society (IES) and the National Kitchen & Bath Association (NKBA) recommend a minimum countertop light level of fifty foot-candles (fc) with all lights blazing. The amount of light is sometimes referred to as lux. Either way, adequate lighting in the kitchen is especially important when using sharp instruments for tasks like chopping vegetables and shredding cheese.

To deliver enough illumination, recessed can lights should be placed directly above the counters, eight to twelve inches from upper-cabinet faces. (If the lights are placed farther out, the cook will block the light.) A quick and effective remedy is to install under-cabinet lighting. Look for LED strips that put out more than 500 lumens per square foot. And don't forget to bring natural light into the kitchen. We design our kitchens with large windows and/or glass patio doors, which provide a connection to the outdoors while adding beautiful ambient daylight. Remember, a dim kitchen can lead to missing fingertips!

We discussed dishware and utensils in Chapter 11, but we'll reiterate here that you should avoid melamine; bisphenols such as BPA, BPS, and BPF; and nonstick chemicals like PTFE and PFOA. Instead, opt for glass, ceramic, stainless steel, and cast iron.

As for appliances, we avoid anything that uses gas, as well as microwave ovens, due to the potential for EMFs. Some great alternatives include a modern pressure cooker, an air fryer, and a good blender with a glass pitcher.

An organized kitchen is key to keeping surfaces clean and sanitary so you can have ample open workspace for your culinary creativity. Maintaining an orderly pantry means regularly purging old and expired food to reduce the potential for bugs and the risk of mold growth. Installing spice racks and accessory drawer organizers is a quick and easy way to add neatness, too. Consider moving countertop appliances to the pantry or under cabinets to free up counter space.

Routine maintenance ensures that food prep areas are kept clean and appliances remain in working order. Make a note to wipe down and deep-clean your refrigerator shelves, drawers, and outer seals every so often, as well as your oven (when it's cool). Scrubbing the cooktop vent filters will remove greasy buildup and keep air flowing properly. It's also a good idea to periodically run your empty dishwasher on the high heat or sanitize cycle with some bleach to disinfect the interior surfaces.

And don't forget your garbage disposal. Fun fact: bacteria love to hang out in the kitchen sink and inside the disposal, including types of bacteria that can make people very sick. Once a week, run ice and hot water through the disposal with some detergent to help break up any grease, food, and grime. To disinfect the rubber stopper parts, scrub them weekly with equal parts water and bleach. Don't forget to scrub and disinfect the surfaces of your sink and faucet, including the areas under the countertop if your sink is undermounted.

Finally, skip the kitchen rag routine and opt for paper towels. Rags rate highly on the bacteria count list, which includes MRSA (a dangerous staph bacteria) and strains of E. coli. If you can't stomach the constant waste, make sure to let your rags dry out well after use, and switch them out at least once a week.

Sources of Kitchen Germs

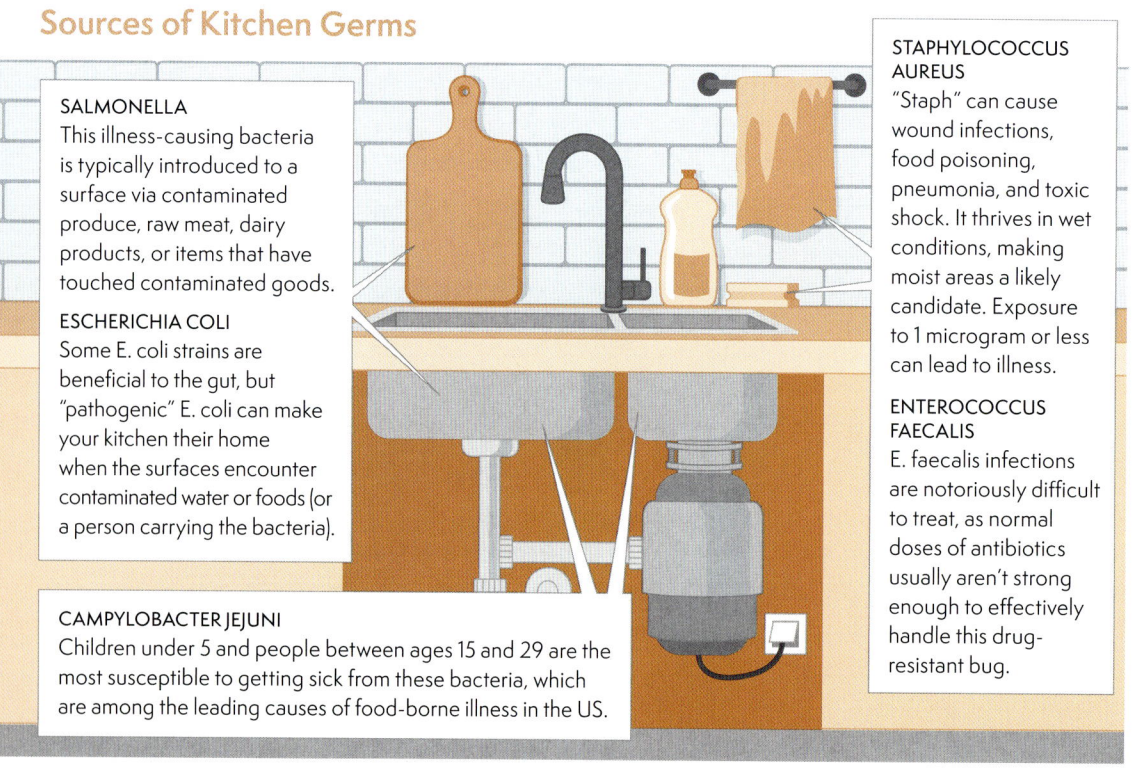

SALMONELLA
This illness-causing bacteria is typically introduced to a surface via contaminated produce, raw meat, dairy products, or items that have touched contaminated goods.

ESCHERICHIA COLI
Some E. coli strains are beneficial to the gut, but "pathogenic" E. coli can make your kitchen their home when the surfaces encounter contaminated water or foods (or a person carrying the bacteria).

CAMPYLOBACTER JEJUNI
Children under 5 and people between ages 15 and 29 are the most susceptible to getting sick from these bacteria, which are among the leading causes of food-borne illness in the US.

STAPHYLOCOCCUS AUREUS
"Staph" can cause wound infections, food poisoning, pneumonia, and toxic shock. It thrives in wet conditions, making moist areas a likely candidate. Exposure to 1 microgram or less can lead to illness.

ENTEROCOCCUS FAECALIS
E. faecalis infections are notoriously difficult to treat, as normal doses of antibiotics usually aren't strong enough to effectively handle this drug-resistant bug.

BEDROOMS AND CLOSETS

As you might expect, our advice for bedrooms starts with the bed. Give ample thought to the mattress you buy, as it can make or break a good night's rest. If you don't have a healthy mattress, it's time to invest in your sleep.

There's a lot of recent questioning about the safety of mattresses, and rightfully so. Mainstream mattresses are treated with arsenic and other dangerous chemical flame retardants. Most manufacturers use soft woods and materials that emit formaldehyde fumes in addition to other noxious odors from petroleum-based foams.

Organic cotton, natural latex, bamboo, and wool are materials to look for when purchasing a mattress. Stay away from soy, as it is often blended with polyurethane foam, which releases noxious VOCs. Keep in mind that any mattress should be replaced after ten years, as mattresses can become twice as heavy with the buildup of sweat and shed skin cells over time.

All the parts of your bed, from the platform to the headboard, should be made of healthy materials. A solid hardwood bed finished with no-VOC paints and sealers is a timeless addition to any bedroom and will last a lifetime. Other healthy materials to look for when purchasing beds and nightstands are metal, acrylic, and stone. Leather headboards are also a great option. Top-grain leather makes a gorgeous statement. (If you have a custom leather furniture manufacturer, ask if they can source softer hides that are minimally treated. Some leather companies even offer vegetable-dyed hides. We recommend forgoing aftermarket upholstery stain treatments.)

Get rid of any bedroom furniture made from MDF or particleboard, which can off-gas formaldehyde for years. The same goes for most engineered wood furnishings.

Avoid mattress protectors that contain PVC; we use protectors made from bamboo or Tencel, which are excellent for moisture wicking and temperature regulation. We also opt for pillows stuffed with natural materials such as organic cotton, bamboo fibers, feathers, or modal. These types of pillows are often labeled "made with hypoallergenic materials."

Don't keep your bed plugged in. Many beds come with electric motors to lift parts of the mattress. Get into a desirable position, then unplug the motor. Otherwise, you'll be sleeping on an EMF hotbed every night. Avoid running electrical wires under the bed or near your head at night, too—this includes cell phone chargers. The best option is to keep your phone at least three feet away from you at night, so put it on the far side of the nightstand.

Other bedroom EMF concerns include ceiling fans (upgrade old fans to ones that have DC motors, which last longer, are quieter, and produce less EMF), alarm clocks (choose the kind that run on batteries and don't connect to other devices via Bluetooth or Wi-Fi), and, if you sleep on the second story, exhaust vents and ceiling fans on the floor below, which have motors whose vibrations and EMFs can disrupt sleep. Make sure to turn off your downstairs fans before bedtime. Minimizing exposure to EMFs during sleep helps ensure healthy circadian rhythms.

Wireless frequencies in the home can also disrupt sleep. Power off your devices that have wireless capabilities and disconnect your Wi-Fi router before bedtime. You'll be surprised by how much better rested you feel even after the first night.

EMF Sources Within a Home

what are circadian rhythms?

In the human body, circadian rhythm describes the twenty-four-hour cycle that runs constantly in the background, synchronized with the master clock in your brain. This cycle ensures that essential biological functions are carried out. The most famous circadian rhythm is the sleep-wake cycle. When in proper balance, this cycle promotes healthy sleep, but when out of balance, it can create all kinds of problems.

A variety of factors and environmental cues can influence the master clock in your brain, including light, exercise, caffeine, and more. Even EMF radiation can affect your sleep cycles by reducing the amount of melatonin your body produces. It should go without saying, but not getting enough consistent and restorative sleep can lead to a slew of health problems. Make sure your bedroom is a sanctuary.

One electrical trick you *can* try is keeping yourself "grounded." We place our grounding sheet on the bottom half of our mattress under the fitted sheet because optimal grounding is said to be done through the feet. Be sure to use an outlet tester before plugging in the ground to ensure the wiring is safe and correct.

Soft, nontoxic sheets are also important for a comfortable night's sleep. Choose materials from nature that wick moisture and are naturally antimicrobial, such as bamboo, modal, viscose, Tencel, and beechwood. These hypoallergenic materials come from renewable resources and help regulate body temperature while sleeping. Organic cotton is also a nice option, but it does not have the same moisture-wicking and temperature-regulation benefits.

Wash your sheets once a week, and periodically toss your pillows in the dryer on high heat or the sanitize cycle to help keep them sanitary (and fluffy!). Another tip is to spray some liquid silver solution or antibacterial hand sanitizer made from water and hypochlorous acid on pillows prior to drying to get them extra clean.

the benefits of grounding

A grounding sheet is a conductive fabric that mimics the effects of being in contact with the surface of the Earth, balancing out natural energies while you sleep. In essence, it returns you to something resembling the natural state of sleeping on the ground outdoors. This works via conductive threads sewn into the fabric along with a connection to the ground port of a standard electrical outlet. Using a grounding sheet may help alleviate sleep deficiencies, inflammation, chronic pain, and more.

Now let's talk about cleaning the bedroom. Ever looked under your bed with a flashlight? It's a great place for dust bunnies and dander to collect. Run a dust mop under your bed once a week to keep dust at bay.

In the closet, wipe down shoe cubbies regularly, as gunk and dirt from the soles of your shoes build up over time. Make sure your closet has good air circulation to keep clothes fresh and reduce any laundry hamper smells.

Speaking of hampers, use one that has holes for ventilation and is easy to wash periodically. (Your clothes are dirty, and so is your hamper!) Also, remember that your belongings should go on shelves, on hooks, on rods, and in drawers. Keep as much as you can off the floor so that it can be cleaned regularly to prevent allergen buildup.

The types of items you store in your closet matter, too. Luggage, purses, and shoes are often made from PVC or polyurethane, which tend to off-gas phthalates (and the accompanying plastic odor) for years. Consider replacing these pieces with ones made from hard plastics, metals, fabrics, and leather, which are cleaner and more substantial materials that won't release VOCs in your closet.

Finally, we all love our pets, but cats and dogs can introduce dirt and germs into the bed after running around outside or using their litter pan. Make sure to clean their paws and wipe off their coats before snoozing with them. Also, use a lint roller each morning to avoid fur buildup on the bedding.

BATHROOMS

One of the best things you can do for your bathrooms is to make sure you're keeping humidity and moisture under control. Upgrade the exhaust fans (and add them near showers if necessary) to ensure proper ventilation in steamy areas. Seal all grout if your shower area is tiled (our preferred nontoxic material for walls and floors) and make sure to maintain waterproofing around the edges of shower escutcheon plates and glass joints with an application of solvent-free silicone.

Check for leaks frequently and address any moisture issues ASAP; leaks not only waste water but also supply a constant source of moisture, which can lead to mold problems. After using the shower, leave the door open and keep the fan running for half an hour. Finally, swap out any fabric floor mats for teak mats, and exchange old towels for naturally antimicrobial and highly absorbent bamboo or Tencel alternatives.

When it comes to cleaning showers and tubs, we recommend handling stubborn soap scum using a damp sponge and a limestone- or feldspar-based cleaning powder. You can use pure vodka, pure 5 percent vinegar, or diluted clear ammonia in a spray bottle as a glass and tile cleaner, too. Just be sure not to get anything acidic on the grout. Splashless bleach works great for disinfecting toilets. (Remember, never mix bleach and vinegar.)

GEEK BOX | ## the benefits of Tencel

Tencel is the brand name of a type of lyocell fabric made by dissolving eucalyptus pulp and then pushing it through small holes to create threads, which are spun into yarn or woven into cloth. This process produces a soft, hypoallergenic fabric that's great for kids and adults, especially those with skin sensitivities.

Tencel requires less energy and water to produce than conventional cotton. It is biodegradable and needs less dye for color than many other fabrics. You can also get Tencel in white without worrying about bleaching because white is its natural color. However, like any product, make sure the item you're looking at wasn't made using a combination of synthetic fibers or harmful chemical additives.

Clean drains periodically with one cup of baking soda followed by one to two cups of white vinegar. Let sit for an hour, then rinse with hot water for sixty seconds. Clean out sink and bathtub overflow holes periodically with cotton swabs dipped in bleach, and replace toilet wands every six months. You can address stubborn clogs by using an enzyme-based chemical drain cleaner (or by calling a licensed plumber). We also encourage the use of trash can liners, but make sure they're unscented; avoid those that smell like recycled detergent bottles.

Check your toilet tanks periodically to make sure the water is clean and there's no sign of mold growth. Be sure to look under the porcelain tank lid, too. To clean the tank, pour bleach into the tank and let it sit for an hour. Mold is hard to remove from toilet tanks once the roots grow into the rough porcelain finish, meaning you may need to replace the tank if you detect stubborn mold issues. Better to keep it clean before mold growth starts.

Sources of Bathroom Germs

Damp towels provide a perfect place for mold and mildew to spread. Make sure to hang your towels where they can dry quickly and change out/wash them often.

You'd be surprised what's on your toothbrush. One study showed that 64 percent of toothbrush holders are contaminated with coliform (an indicator of potential fecal contamination).[1]

Germs can transfer from your hands to the sink while washing. In fact, most of the bacteria on these surfaces comes from human contact rather than fecal bacteria.

The damp environment of the toilet tank can be a breeding ground for mold, so make sure to clean it with bleach occasionally.

Surprisingly, the average toilet isn't as dirty as you might think. That's because most people are highly aware of its dirtiness and try to keep it clean. Just remember to put the lid down before you flush.

Cleaning your body is just as important as cleaning the bathroom surfaces. We use unscented and naturally scented soaps and shampoos and opt for washcloths instead of loofahs, which can be breeding grounds for germs. Launder towels and washcloths weekly, and let your towels hang dry after use in a well-ventilated area to avoid mildew. And speaking of soap, one of the most common places for bacteria to breed in a bathroom is on the soap dispenser. Clean it regularly, too; motion-activated soap dispensers can help cut back on the number of times the dispenser is touched.

For shower curtains, find a brand that does not contain PVC; look for PVA or PVE liners and curtains made from natural materials such as hemp, cotton, flax, and cork. To enhance comfort as well as health, add a portable infrared heater on casters to quickly warm a cold bathroom.

Even if you have a whole-home water purification system, it can't hurt to install purification at the point of use on showerheads and bathtub faucets. If you live in an older house, you never know what might be leaching into the water from your pipes.

BABIES + KIDS

Infants are prone to dry skin, booboos, and allergic rashes. Baby balms made of organic plant squalene and naturally derived oils that are fragrance-free are the safest and most effective options. Fragrance-free wipes made of 99 percent water are also the way to go. We all want to avoid those painful, insidious diaper rashes.

Diapers made from plant-based materials with zero chlorine or fragrance are worth the extra few cents. Not only are they better for your baby's health, but you can rest easy knowing your baby is not absorbing chemicals through their skin 24/7. But some natural substances can cause reactions, too. Avoid baby skincare products that contain shea butter, lanolin, and zinc oxide, which may be sensitizing.

Sadly, most children's clothing is treated with fire retardants that have been linked to cancer, thyroid problems, reproductive system issues, and sensitization to allergens. Look for kids' clothing brands that are open and honest about their fabrics and what goes into them. OEKO-TEX is a good label to look out for when making safe and sustainable purchasing decisions. Although they tend to be more expensive, several clothing lines carry cute kids' attire made of bamboo rayon. The difference in softness is remarkable, plus these clothes are usually not treated with harmful chemical flame retardants.

 GEEK BOX amazing modal

Modal is a semi-synthetic cellulose fabric, more durable than viscose rayon but still breathable, absorbent, and soft with impressive resistance to shrinkage and pilling. It is derived from beechwood, meaning it has the potential for sustainable manufacturing, too. This is why it's so popular for underwear and activewear, but it doesn't explain why we love it for a healthy home.

Not all modal is perfect, but fewer harmful chemicals tend to be used in the production of modal than many alternatives, and some manufacturers have started to use machines that trap the chemicals before they make their way into the ecosystem, further reducing harm. Modal also happens to be wonderfully hypoallergenic and quick to dry, meaning it won't harbor potentially harmful bacteria.

Choose children's bedding made from bamboo, Tencel, modal, or beechwood. These natural and healthy options will help keep your babies cool on hot nights and warm on cool nights. Look for an organic, no-VOC mattress, too. Some brands have corn-based waterproofing applications for infant sleep pads, which are a wonderful alternative to polyurethane or PVC. Organic cotton (or any of the materials mentioned above) is a great, soft, chemical-free choice for pillows. To find a crib, follow the directions for sourcing a bed on page 272.

Consider switching to bamboo towels and washcloths for little ones. Bamboo holds up longer than cotton, is better at absorbing water, is a bit more stain resistant, and naturally inhibits bacterial surface growth and odors.

Taking tumbles is a rite of passage for babies and toddlers, and of course parents want to ensure that the flooring is forgiving. We recommend ceramic tile, which is cold and hard, but rugs made of viscose and wool are excellent options if not treated with pesticides, fire retardants, or stain repellents. Plus, they're soft, plush, and beautiful accent pieces. We haven't had much luck sourcing nontoxic interlocking foam flooring systems; even closed-cell foam seems to off-gas. If you're looking for extra padding, consider putting a layer of cork beneath the rug. (See Chapter 18 for more about finding the best rugs.)

Let's switch gears and look at baby- and toddler-proof dishware, where functionality is the name of the game. Most dishes and sippy cups are made from plastic that contains BPA, BPS, and/or BPF, but you want to look for options made from stainless steel, silicone, or bamboo composites. Glass is the best option for storing and cooking kids' food.

Beverage manufacturers have caught on to the healthy baby craze and have released lines of baby waters in large jugs, but milk jug plastic can leach into the water, especially if it is slightly acidic. We only drink spring water from glass bottles, which is naturally alkaline with balanced minerals straight from the earth.

Don't heat food in plastic, foil, or Styrofoam. (This goes for your food, too, Mom and Dad.) Instead, use glass, ceramic, porcelain, or stainless steel for cooking and storage, as these materials won't leach and are safe to heat. Certain silicone containers are also labeled heat-safe, and all food-grade silicone is perfect for storing foods at room temperature or in the refrigerator.

If you're concerned about proper nutrition, you can even make your own baby food. Take some of what Mom and Dad are having for dinner and puree it for the baby. We found a food steamer and blender all-in-one that is an especially valuable time-saver (with fewer dirty dishes as a bonus).

But be careful about allergies, which are more prevalent in children now than ever before.

Nuts are a common allergen for kids. Be sure to read labels carefully because more and more food manufacturers are using nut ingredients such as almond flour and cashew butter. Don't assume all plant-based products are OK. Shea butter also comes from a nut and is found in many topical lotions and cosmetics. We've found that some popular lotions labeled hypoallergenic contain macadamia nut oils. Keep in mind that sometimes plant-based ingredients are labeled by a plant's scientific name, so do your homework to make sure it's safe for any family members with allergies.

Latex is another common allergy that many people don't realize can be triggered by certain foods. Ripe avocados, ripe bananas, tomatoes, and some uncooked veggies have similar proteins to latex that can cause reactions in sensitive individuals. Latex allergies can be quite serious. Latex is a contact allergen, so use caution when buying carpeting and rugs that may contain it. Make sure your baby bottle nipples are made of pure silicone, and buy bandages that are labeled latex-free.

food allergies and kids

Food allergies are no laughing matter. Allergies can appear in children as young as newborns, which can affect breastfeeding for many new moms, as the proteins from foods stay in their milk for three to six hours after eating. The most common food allergies are dairy (as in cow's milk), eggs, nuts, soy, and wheat. The resulting symptoms can range from diarrhea and colic to, in severe cases, death. According to Johns Hopkins Medicine, nearly 5 percent of children under age five have food allergies.[2]

So, what can you do about it? If your child has food allergies, one of the best things you can do is to eat a rotation diet. This is a system of managing food allergies that involves eating biologically similar foods and then waiting at least four days before eating them again. Such a diet can prevent new food allergies from developing, help you identify previously unknown allergies, and allow kids to eat a greater variety of foods, especially ones they might have a reaction to if they ate them more often. For example, if you eat chicken on Monday, you should not eat it again until Friday. In addition, eggs and other foods from the same "family" should be rotated on those days. For more information about a rotation diet, speak with your doctor.

Ask your nutritionist and pediatrician about vitamins and other supplements for your kids. Much of today's farmland is devoid of essential minerals such as magnesium. The resulting foods are lacking essential vitamins, minerals, and enzymes. Supplementation has basically become the standard way to reach healthy nutrition levels in modern living. Vitamin gummies are a fun treat for everyone!

Moving on to general safety, remember to cover those electrical outlets, of course. But training is often a better solution. Explaining and ingraining rules such as "don't open cabinets and drawers" may take more time up front, but you'll be able to use your kitchen instead of working around tons of child locks. A less stressful and more functional space is conducive to good health and well-being.

Baby monitors that run via radio waves are a much better option than Bluetooth or Wi-Fi-enabled baby monitors. The latter emit radiation 24/7, which is especially concerning for a child during critical stages of body and brain development. In that same vein, forget the wireless sock and diaper monitors. Very few parents need to constantly monitor their child's heart rate while subjecting the kid to EMFs all day and night. Check on your child the old-fashioned way. Everything will be fine.

Also, take a closer look at your kids' toys to see what they are made from. Solid wood toys are fabulous and can take a beating. Just make sure the wood isn't an engineered or pressed wood with layers held together by adhesives. Even if the glue is formaldehyde-free, the soft wood is not. Go for toys made from solid hardwoods, hard plastics, silicone, and lightweight metal. Kids chew on everything, so make sure the paint on their toys is nontoxic as well.

Add stylish storage solutions to your child's bedroom, bathroom, and playroom with woven baskets crafted out of reeds, rattan, wicker, jute, or cotton/rayon string. As previously mentioned, we generally avoid water hyacinth and seagrass for baskets or accessories because these materials tend to promote mold growth.

Thinking about painting the nursery or doing a mural in your kids' bedroom? See Chapter 14 for important information on selecting the right kinds of nontoxic paints and finishes.

LAUNDRY

When it comes to doing laundry, keep a plant-based detergent handy for use on everything from baby clothes to sheets to adult apparel. (The best options are usually referred to as "plant-based," "natural," and "free and clear." These detergents contain zero fragrances, dyes, and substances known to cause allergic reactions.) Detergents made from soap nuts are gaining popularity and produce amazing cleaning results. (Contrary to the name, soap nuts are actually berries and are perfectly safe for people with nut allergies.) Pro tip: add a half cup of clear ammonia to each load to preserve color on clothes and remove stubborn oil buildup.

Skip dryer sheets. They're not necessary and tend to add a waxy residue to everything being laundered. Instead, use 100 percent wool dryer balls to reduce drying time by up to 25 percent and keep wrinkles at bay. Hang clothes to dry in an area that won't get damaged from a few drops of water, such as over tile, a bathtub, or a solid countertop.

Most natural, plant-based clothing materials (cotton, hemp, lyocell/Tencel, beechwood, bamboo, modal, etc.) last significantly longer if hung to dry. If you fold shirts, pants, and dresses made of natural materials in half over a hanger, you can prevent stretching from the wet weight. Take care to secure long sleeves.

OUR TOP THREE LAUNDRY HACKS

Here are three of our favorite safe, natural tricks to use in the laundry room:

- Lemon juice is a natural bleaching agent. It can help remove stains from whites while going easy on your skin.

- Hydrogen peroxide is great at getting rid of protein-based stains like blood and other bodily fluids.

- Adding a bit of distilled white vinegar to a load of wash can make clothing more vibrant, kill mildew, and even help remove a variety of stains. Just make sure never to mix vinegar and bleach. Note that for natural fabrics such as bamboo, a very small amount of vinegar will get the job done.

Scrub your washing machine drum with a non-scratching abrasive cleaner and a scrub sponge. Then run the cleaning cycle with hot water and bleach. This is very important and often overlooked, as dirty clothes will deposit grime in your washer. Pour bleach or hydrogen peroxide down your washer drain, located in the wall behind the unit, periodically to prevent grime buildup.

Install an exhaust vent in your laundry room. Even better, upgrade to one with a humidity-sensing fan that kicks on during warmer laundry cycles. This can go a long way toward preventing a mold problem before it starts.

Ensure that your dryer vent is clear of debris and lint, as clogged dryer vents are a fire hazard. Also make sure the vent exits on the side of the home (not underground), and double-check that the vent has a bug screen and a functioning damper. Otherwise, you'll wind up with bugs, pollen, and dust in your dry clothes.

Check whether your laundry vent pipe is PVC or metal. PVC dryer vent pipe was recently deemed a fire hazard after it was found that static electricity could cause lint to spark and catch fire. Ask your HVAC professional to replace any PVC exhaust ventilation with metal flex duct.

If possible, avoid doing laundry while you sleep. The washer and dryer are not only loud but also can emit high amounts of EMF inside a home. Also, avoid running your washing machine on the high spin speed cycle, as it is designed to compound the motor's power from within, which creates a large and powerful EMF field. The low or medium spin speed is a better option that still gets the job done.

GENERAL CLEANING TIPS

Always keep a jug of clear (not lemon-scented) ammonia on hand. It's safe to dilute and use in well-ventilated areas and removes grease and grime from clothing, ovens, glass, mirrors, and many other hard surfaces. (Avoid inhaling ammonia fumes while cleaning.) Pro tip: a 1:1 mixture of clear ammonia and water in a spray bottle will remove mycotoxins from most hard surfaces and remove oily, waxy residues on hairbrushes and other hard-to-clean surfaces. Use it only in a well-ventilated location.

Remember, a clean home is a healthy home, so we recommend keeping to a cleaning schedule. This includes vacuuming flooring and rugs with a HEPA-rated vacuum cleaner twice a week and dusting once a week with fragrance-free duster pads. Take care to avoid purchasing duster pads that have detergent brand chemical additives. We did this on accident years ago and spread a strong laundry detergent fragrance across a whole section of the house.

Clean toilet bowls, showers, sinks, bathtubs, and other hard surfaces weekly. Wipe down solid surfaces such as kitchen countertops on a daily basis. Scrub the grime off the laundry washer drum once a month. Wash your sheets, towels, and fabric mats weekly and your bedspreads at least monthly. We recommend breaking up the weekly cleaning tasks, doing different ones on different days, so that you're not stuck trying to get it all done on Saturday.

WHAT'S IN THE AIR?

When it comes to healthy air, make sure your home has adequate ventilation and filtration. This means opening windows occasionally and using MERV-rated filters of 13 and up in your HVAC. But keep in mind that no amount of filtration will help if the materials in your home are constantly off-gassing.

As such, follow the guidelines elsewhere in this book, but also forgo scented plug-ins and fragrance sticks. Synthetic fragrances are usually derived from petroleum-based products that are irritating to the body and can lead to allergic sensitization. Instead, spark up the ambiance with candles made from naturally derived waxes. You can't go wrong with an unscented candle made of soy, tallow, or beeswax. If you're looking for a relaxing aroma, some natural candle makers also have organic essential oil scents.

You should also think about what's under your home, because one of the leading causes of lung cancer is radon, an odorless, tasteless airborne gas that is a by-product of natural rock decay, which we discussed earlier in the book. This gas seeps through cracks in a home's foundation and into plumbing penetrations.

Radon levels are difficult to determine based on location alone; one house may have high amounts while the house next door has none. Quick, inexpensive radon tests can be ordered online. Whether or not you find a problem that needs to be addressed, you'll be glad you did it.

GEEK BOX | the surprising benefits of tallow

Tallow, or the fat rendered from beef, is found in soaps and is even available as a topical body conditioning oil. (Healthy fat equals healthy skin!) Plus, it can be used to condition leather furniture without making it stretch.

But perhaps its best use is for healthy, natural cooking. Beef tallow contains good-for-you fats and has a high smoke point, meaning it takes longer to burn, retaining more of its nutrients and not imparting harmful carcinogens to foods. There's also evidence that the conjugated linoleic acid and other nutrients tallow contains may have anti-inflammatory and immune-supporting properties. Make sure the products you use for cooking and skincare are sourced responsibly from organic cattle.

THE LOWDOWN ON ESSENTIAL OILS

Essential oils are concentrated plant extracts obtained through mechanical pressing or steam distillation that retain the natural smell and flavor of their source. As such, these liquids can be incredibly aromatic and have a variety of applications, ranging from perfumes to medicinal products (thanks to their antimicrobial and antioxidant properties). While most essential oils are considered safe, some can be irritating and produce allergies with prolonged airborne exposure, so use them with caution.

ESSENTIAL OIL USES

STRESSED
Lavender   Eucalyptus

TIRED
Orange   Peppermint   Lemongrass

INSOMNIA
Lavender   Marjoram   Chamomile

SORE THROAT
Oregano   Lemon   Thyme

HEADACHES
Orange   Ylang-ylang   Marjoram

DRY SKIN
Lavender   Tea tree

DISTRACTED
Peppermint   Lemon

There are many hidden sources of VOCs in a home, but a pair you might not suspect are newspapers and magazines. The inks are made using xylene, which is a harmful carcinogenic aromatic hydrocarbon. Reading the news is great, but we suggest doing it on the patio and throwing away the publications outside when finished. (Markers are also chock-full of solvents such as xylene, so throw out the stinky markers and opt for nontoxic versions that come in both washable and permanent ink.)

Likewise, some people love to collect old books, but old books are just that: old, dusty, and likely coated in mycotoxins after being kept in moldy environments. Unless you're absolutely sure where an old book came from, it's best to toss it.

If you're a fan of indoor plants, skip the soil and go for hydroponic planters. Or, better yet, higher-end fake plants are usually safe alternatives, and so are silk plants that are not treated with pesticides. Some retailers now offer dried plants that are cured with formaldehyde-free biobased formulas, another excellent maintenance-free option.

On the topic of general safety, check your smoke alarm batteries annually. Most alarms have a light that starts to blink when the batteries are getting low, which means it's time to replace them. If you use natural gas appliances (which, remember, we don't recommend), make sure your home has a properly functioning carbon monoxide detector as well.

YARD CARE

It's alarming how many of the pesticides used in today's conventional farming practices are toxic. If you cover your own home in such pesticides, you're asking for major health problems. That's why we recommend keeping bugs at bay the natural way.

Diatomaceous earth is a great broad-spectrum natural pesticide that can be used indoors and out. This naturally mined silica clay powder is chemical-free and even safe for use around kids and pets. The only place you shouldn't put it is around vents, as inhaling it can irritate the lungs. Neem oil and thyme oil are both effective insecticides for outdoor use, and copper sulfate–based sprays are a good option for some fungal plant diseases. You can even try a simple spray bottle concoction of water, a few drops of plant-based dish soap, and a tablespoon of hot sauce to keep grasshoppers and other bugs off your herbs. Guess they don't like spicy foods!

Spreading corn gluten meal over grass in early spring will help deter weeds from popping up later in the season. This organic pre-emergent weed control method is only successful if watered in well after application, followed by one to two days of no water (check the weather forecast). Multiple applications six to eight weeks apart is even better.

 GEEK BOX | ## the health effects of pesticides

Pesticides are designed to be toxic. Otherwise, they wouldn't dispatch the pests they target. Sadly, they can also be hazardous to humans and pets, especially in large doses or over long periods.

The symptoms of pesticide poisoning are wide and varied, ranging from mild skin irritation to death, and are typically classified as either topical or systemic. (Different classes of chemicals produce different symptoms, and individuals vary in their sensitivity levels.)

Topical effects usually develop at the site where you encountered the pesticide and may include red skin, rashes, coughing, and wheezing, while systemic issues occur in other areas of the body—a sign that the pesticide has been absorbed and distributed. Systemic effects may include nausea, fatigue, headaches, tremors, and other disorders.

Acute symptoms such as pinpoint pupils often occur right after exposure but could appear later. If you live in the US and start having symptoms you suspect might be related to pesticide exposure, call the National Poison Center at 800-222-1222.

Skip the Wi-Fi enabled sprinklers; use a wired rain gauge instead to avoid overwatering your lawn. Creating a one- to two-foot border of small rock (half-inch to two-inch stones) around your home can also help with drainage around the foundation, plus keep dirt and moisture away from the edges of the home.

Avoid smart-home pool equipment that responds to a Wi-Fi signal. You can easily get the same benefits by pre-programming the hot tub to sizzle up at a certain time of day so it's hot and ready when you get home.

WINDOWS + DOORS

Your home's entrances and exits are important for a variety of reasons, from insulation to moisture protection.

It all starts with the welcome mat. Most of us don't think twice about what we're bringing in when we walk through the front door or garage entry. You stamp your feet on the mat to remove the dust from your shoes and then enter.

In fact, that welcome mat is one of the dirtiest spots in the house. Shoe soles can carry traces of coliforms and fecal bacteria, and the welcome mat is a great place for these bacteria to grow. Washing your mat using a high-powered hose with some all-natural detergent will help keep it clean, and spraying it once a week with a nontoxic disinfectant spray will keep the microbe count in check. Plus, making it a household policy to leave your shoes at the door is one of the best ways to keep the floors clean throughout your home.

When it comes to windows, consider replacing older single-pane windows with double- or triple-pane windows, which provide much better insulation and sound protection. Add awnings over windows that are not adequately protected from wind and water by the roof overhang.

Next, check the flashing on the exteriors of the windows. Repair any areas that look worn or cracked. On the window interiors, note any spots where condensation is building up and consider replacing them with window models that have a true thermal break. Fogging is a sign that the seals on your double-pane windows have failed.

Finally, install adequate weather stripping around all exterior doors. See the windows and doors sections in Chapter 9 for more details.

WATER HEATERS

The average life span of a water heater is between ten and twenty years. That might seem like a huge range, but it can vary depending on the type of heater and whether you're using gas versus electric or tank versus tankless.

First, let's talk gas versus electric. Again, we don't recommend gas appliances inside the building envelope, as gas heaters can release raw natural gas and carbon monoxide into the house. Those who are unknowingly sensitive to natural gas may experience mood swings or depression due to natural gas exposure. A gas water heater should be installed in an insulated closet or garage on the exterior of the home (outside the building envelope).

If you do go for a gas heater, know that they tend to cost a little more up front but a little less to operate. They are also usually larger and more efficient than their electric counterparts, but electric units require less maintenance and will likely last a bit longer—ten to fifteen years if properly maintained. Installing a whole-home water purification system can add to your water heater's life span.

Tankless units, however, beat them both with an average life span of over twenty years. These units use high-power burners to heat water quickly, ensuring you have an on-demand supply whenever you need it. The downside is higher upfront costs, but tankless heaters are usually more energy efficient and often use less water to boot.

Tankless Versus Tank Water Heaters

TANK:

- Big and bulky
- 24/7 electricity use
- Indoors
- Shorter installation time
- Limited hot water supply
- Water sits in tank
- Low upfront costs
- High lifetime costs

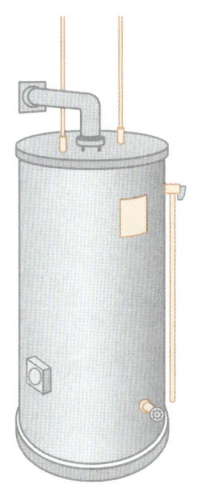

TANKLESS:

- Suitcase-sized
- As-needed electricity use
- Indoors or outdoors
- Longer installation time
- Unlimited hot water supply
- Fresh, clean water
- High upfront costs
- Low lifetime costs

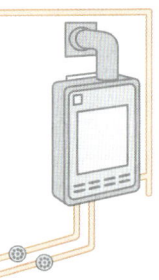

One issue you might encounter with a tank water heater is mineral buildup, which may be more likely depending on the water quality where you live. You can tell this is happening if you hear clanking or rattling inside your unit, the water temperatures are inconsistent, or the water comes out rust-colored. Sediment buildup will shorten the life span of your heating unit, so it's recommended that you flush your water heater at least once per year. Doing so will help prevent the problems that sediment can bring about over time, such as coating the heating filaments within. If you have hard water (or water that contains more minerals), you may want to flush your tank more often. Be aware that large amounts of sediment buildup may require a professional to unclog the flush valve.

Check to make sure your water heater has a moisture-sensing shutoff switch in the drain pan. An overflow could become a catastrophic flooding event, so it's important to have moisture-sensing equipment in place to shut off the appliance if it breaks. Not all builders are careful about installing simple yet critical components such as drain pans and moisture-sensing shutoff switches. It's important to make sure your water heater has a drain pan below and that the drain is functioning properly. It's a good idea to call a licensed plumber to get these installed if you don't have them.

Peek inside your water heater closet or attic to see if the area needs to be vacuumed or cleaned. Many water heater/maintenance closets are unfinished, which leaves unhealthy construction materials exposed in many homes that weren't built to our health standards. Consider sealing exposed materials to prevent off-gassing.

Also be sure to winterize any tankless water heaters mounted on the exterior of a home during the winter months. They can freeze and burst during a cold snap. The good news is that this would typically result in a leak outside the home, meaning fewer internal mold issues, but the bad news is that your water heater will need to be replaced. There are insulated jackets for exterior heaters. Ask your plumber if additional options are available in your region.

GARAGE

Finish out your garage. Even though it's not a space for your living room couch, it is often visited and stores a lot of frequently used items. Keeping windows cracked in the garage will allow for ventilation and air flow when it's not raining. (Consider adding cross-ventilation vents in the garage walls if there are no windows.)

Making your garage a more livable space means storing all gasoline, diesel fuel, paint, and chemical products in a storage shed away from the home (if the garage is attached to the home). It also means washing your garage floor with a heavy-duty nontoxic detergent every six months. Dirt, dust, and bugs are just a few of the many things that like to hang out on a garage's concrete floor.

Does that mean you can't utilize your garage space for storage? Of course not! Powder-coated metal or finished-out built-in cabinets are excellent storage solutions. Keeping items off the ground as much as possible will help keep a garage swept and tidy.

Finishing the garage floor with a 100 percent solid epoxy is acceptable for a healthy home—if the garage is located outside of the conditioned space. If you do so, make sure the garage is well ventilated and the epoxy has had time to cure before you use the space. Epoxy is especially toxic before curing, but once cured, there should be only a tiny amount of off-gassing, if any.

Using a penetrating radon sealer or sodium silicate–based sealer on the concrete floor is a great idea for adding another layer of moisture and seepage protection.

ROOF

We recommend metal roofs for a variety of reasons, but that doesn't mean they're totally maintenance-free. You should replace the endcaps on a metal roof every few years and check the integrity of the seams every one or two years. If a section is damaged, don't try to repair it; replace any cracked, split, broken, or missing sections ASAP.

Ensure your gutters are functioning properly. This includes making sure they aren't clogged with leaves, holding water, or dripping or leaking at the seams. If you spot any issues, hire a gutter professional who is trained to scale roofs and climb long ladders. Ask the pro to install gutter screens if leaves are a problem. Also make sure your downspouts are directing water away from your home's foundation.

If you have a fireplace that uses gas or propane, consider switching to a direct-vent unit that fully encloses any combustion fumes. Regardless of fireplace type, have it serviced annually by a professional who is familiar with the type you have. Also have your chimney cleaned annually, as soot is a major pollutant and fire hazard, and check the flashing and waterproofing, because chimneys are hot spots for leaks.

SIDING + EXTERIOR

Keep an eye on the outside of your home and fix minor issues before they become major expenses. This includes checking your roof for missing shingles and tiles and inspecting outside walls for cracks in mortar and stucco, which can develop as temperatures fluctuate over time.

Make sure the land surrounding your home is sloped to properly drain rainwater away from the foundation. If not, consider installing a drainage solution like a French drain or regrading the land to create swales that route water away. Depending on where the water is coming from, building a pony wall to divert water shed from a neighboring property is an option. And sometimes, hiring a professional to install a series of area drains is the best way to fix drainage problems. Talk to your contractor about sizing the pipe to tie in your gutter system as well.

Avoid using penetrating oil protection on exterior lumber. While these products protect wood from UV damage, the oil-based solution doesn't create an effective seal around the lumber and will continue to off-gas petroleum solvent fumes, especially during the warmer months. Instead, choose a product that creates a seal around the lumber in addition to offering UV protection with the infused pigment. This recommendation also applies to decks.

Stay away from fabrics on the exterior of the home. Accessories such as upholstered overhangs and fabric awnings will stain and eventually become a hotbed for mold and mildew. Solid and semisolid surfaces are preferred.

key points	▸ The kitchen is an important room, and it's high on our priority list. Familiarize yourself with the kitchen work triangle and common places where germs tend to collect.

key points

▸ The kitchen is an important room, and it's high on our priority list. Familiarize yourself with the kitchen work triangle and common places where germs tend to collect.

▸ Remember to go for laundry detergents labeled "plant-based," "natural," and "free and clear" while avoiding dryer sheets and other harmful laundry additives that you'll end up wearing all day.

▸ Keep up with your home's exterior and yard maintenance using nontoxic treatments. For example, diatomaceous earth is a great natural pesticide.

▸ Windows and doors are the breakpoints in your home's healthy shell. Make sure they're in good repair and leave harmful substances (like those you may track in on your shoes) outside.

▸ Finish out your garage and store hazardous chemicals in an outdoor storage shed away from your living areas and workspaces.

If your home has a mold problem, you won't want to miss Chapter 20, which is all about mold remediation.

got mold?
steps for safe
removal
and regaining
your health

It seems like only yesterday that Jen discovered the black mold problem in her Dallas apartment. In reality, it was nearly ten years ago, but it still proved firsthand that living in a home, attending a school, or working in an office building with an indoor mold problem can lead to catastrophic consequences for your life and health.

Mold is in the air around us all the time. It may fluctuate at different times of the year depending on where you live, but there is largely nothing you can do about it. Luckily, these background mold counts are rarely high enough to cause illness. The true problem is an indoor environment with an active mold infestation. This means mold has taken hold and is growing.

To thrive, mold requires a food source and water. A lot of building products and contaminants within a home can serve as the food source, including drywall, paper, lumber, and dust. Moisture can come from a leak, improper ventilation, areas in the home that continually create condensation, or high indoor humidity.

If you feel unwell and can't figure out why, you might have a hidden mold problem, which is when you can't smell it or locate the source. But even if you can't see it, a mold problem is serious. We cannot stress enough how important it is to take immediate action, even if you're not feeling any ill effects yet from chronic exposure. If building a totally new house isn't an option, here's our advice on how to handle an existing mold problem.

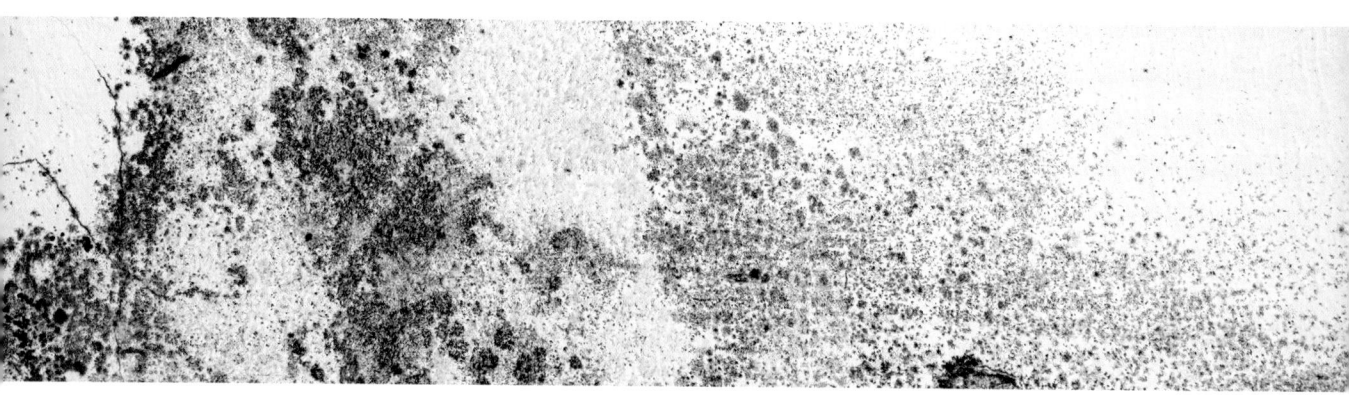

STEP 1: MAKE PLANS TO RELOCATE AND FIND AN EXPERT

Relocating temporarily while a mold remediation process is planned and performed is an important step. You want to get out of that unhealthy environment as quickly as possible.

Next, hire an environmental expert to conduct mold testing in your home and determine what types of mold are present and where. You can also perform your own testing and send the samples to a mycology lab.

Healthier Homes by JS2
July 1, 2021 · 🌐

4 little spores - and that's it! Mold air test results from our latest new build, just completed this month. The outdoor air control was 1800 so I'd say we passed with flying colors. Here's to #HealthyHomebuilding!!

Detailed Mold Report   (WATER-INDICATING FUNGI, IF PRESENT, ARE SHOWN BELOW IN RED)

Analysis Method	Air Analysis			Air Analysis			Air Analysis			Intentionally Blank
Lab Sample #	52464768-1			52464768-2			52464768-3			
Sample Identification	13508680			13426179			13426264			
Sample Location	OUTSIDE			LIVING / KITCHEN			UPSTAIRS HALLWAY			
Sample Type / Metric	Breeze ST/150L			Breeze ST/150L			Breeze ST/150L			
Analysis Date	Thu June 24, 2021			Thu June 24, 2021			Thu June 24, 2021			
Determination	CONTROL			NORMAL			NORMAL			
Fungal Types Identified	Raw Count	Spores/m³	% of Total	Raw Count	Spores/m³	% of Total	Raw Count	Spores/m³	% of Total	
**Non-Problem Fungi										
Alternaria	22	147	1	1	7	35	---	---	---	
Ascospores	215	1,441	11	---	---	---	---	---	---	
Basidiospores	129	864	7	---	---	---	---	---	---	
Cercospora	3	20	<1	---	---	---	---	---	---	
Chaetomium	1	7	<1	---	---	---	---	---	---	
Cladosporium	1,462	9,795	79	2	13	65	---	---	---	
Epicoccum	1	7	<1	---	---	---	---	---	---	
Pithomyces	2	13	<1	---	---	---	---	---	---	
Rusts	3	20	<1	---	---	---	---	---	---	
Smut/Myxomycetes	3	20	<1	---	---	---	---	---	---	
Unclassified Pigmented Spores	---	---	---	---	---	---	1	7	100	
Total Spore Count#	1,800	12,000	100	3	20	100	1	7	100	

This social media post showcases the result of our Healthy Building standards, which involve filtering all air as it comes into the house. As a result, only four mold spores were found in an indoor air test. Keep in mind that it's virtually impossible to reduce spore counts to zero due to ever-present mold spores in the air outside.

PERFORMING YOUR OWN MOLD TEST

First, look around. Check for surface mold and signs of interior or exterior leaks. Inspect your ductwork, as mold is common in HVAC systems. Any dark and damp area could be the source of your problem.

Keep in mind that the EPA doesn't recommend routine mold sampling. Their logic is that if it's all going to be removed anyway, identifying the species of mold you have is unnecessary. Plus, there are no federal standards for mold limits. That said, even the EPA admits testing can be useful in certain circumstances, so if you want to perform a test, go for it. Just keep in mind that the EPA recommends working with a mold remediation professional.

Mold test kits are widely available at home improvement stores and online. Be aware that different types of tests serve different purposes; some determine whether mold is present on surfaces, while others detect mold spores in the air. The testing process can take several days, after which you send the samples to a lab, which will analyze the results.

One example is a mold air test with an outdoor control, which is the gold standard because the outdoor control ensures accuracy. Another new industry test is the environmental relative moldiness index (ERMI) test, which uses dust samples to quantify thirty-six molds. Surface swab mycotoxin testing is a relatively new type. We recommend testing areas that are not likely to be cross-contaminated with mycotoxins from outside sources, such as from your shoes or transferred from clothing onto the floor (because of shoes), couches, chairs, beds, and so on.

STEP 2: CONTACT YOUR INSURANCE COMPANY

See what kind of coverage may apply to your situation through your homeowners' or renters' policy. Find out the insurance company's next steps before beginning any type of remediation or cleanup. The adjuster may be required to visit the site to document the damage.

It is worth noting that if the source of the moisture was a leak in the exterior of the home, your insurance company may deny the claim unless you have flood insurance. Most homeowners' and renters' policies cover a mold claim only if the water issue stemmed from an indoor leak. For those of us who live nowhere near an ocean, other body of water, or flood plain, this news can be somewhat shocking. Nonetheless, it may be worthwhile to purchase flood insurance no matter where you live.

Policies almost always have a maximum out-of-pocket amount. Pay attention to what your insurance will cover and weigh that amount against your deductible. Extensive damage can be costly, so keep your wits about you. In extreme cases, it may be best to start over and buy all new belongings. You'll avoid the costs of storing, moving, and cleaning an entire home's worth of items. Also, some household items like upholstered furniture and rugs are nearly impossible to get completely clean when it comes to mycotoxins.

GEEK BOX | the efficacy of fogging systems

Fogging systems usually use grapefruit seed extract, benzalkonium chloride (when the home is vacant), and/or essential oils to attempt to neutralize mold. These substances may be safe, but none of them removes mycotoxins, and the jury is still out on whether these treatments are effective at killing mold colonies and their spores.

Since mold is often behind walls and other materials in a home, there's no way to get these eradication chemicals into the areas where they can be effective. Mold infestations grow roots into materials, so removing them completely means ripping out everything it's growing on (and even the surrounding areas that do not appear to be contaminated yet). Ozone machines are also great to run (while no one is home) to keep the air in a house clean, but even ozone will not kill or remove a mold infestation.

STEP 3: HIRE A REPUTABLE, LICENSED MOLD REMEDIATION COMPANY

This company will assess the damage and provide a scope of work and a quote for remediation services. Ask if they plan to remove and clean up both the mold and the mycotoxins. Some companies now claim they can do both.

Find out what types of chemicals they plan to use inside your home, and make sure you're OK with what they're using and where they're going to use it.

Make sure they're planning to remove all wet and/or moldy sheetrock, wood, flooring, subflooring, and insulation. It's best to remove and dispose of all carpet as well and replace it with solid-surface flooring. Rugs should be replaced. Request that the team use heavy-duty contractor bags for disposal and that they tightly seal the bags and place them outside. Have the bags hauled away as soon as possible.

The company should use a large HEPA vacuum to go over every square inch of the remediated area and ensure dust and spores are removed. Vacuum brushes should be sterilized and cleaned before work begins on your home.

Depending on the size and severity of the mold infestation, it may be better to remove and dispose of all HVAC interior motors and ducting (if it's flexible ducting—metal ducting can be cleaned without being removed or replaced). Repainting walls, ceilings, and cabinets with a nontoxic paint that effectively blocks chemical emissions can make a big difference in any mold remediation project.

If the mold is confined to a small area and hasn't been there long, ask the remediation team to seal off those areas and the HVAC registers from the rest of the house before starting. If mold problems went unnoticed for a while and were allowed to spread, the entire home will need to be cleaned.

STEP 4: HIRE A REPUTABLE BUILDER/GENERAL CONTRACTOR

BEWARE OF HOME INSPECTORS WITHOUT PROPER CREDENTIALS

We've encountered a handful of individuals who falsely claimed to be professional home inspectors. While they may claim to specialize in finding environmental problems such as sources of mold or EMFs, they are simply preying on families in dire straits.

There are a few institutes we are aware of that train and certify real professionals, such as Building Biologists. It's important to verify credentials and make sure that any professional you contract is abiding by state regulations. You should also ask for a quote up front and negotiate a contract and pricing prior to starting any work.

Get a builder or GC involved before the remediation starts. If your state requires builders to be licensed, make sure they are. It's best to have teams working in tandem rather than hiring a company to do both the remediation and build back (which refers to the process of rebuilding parts of an existing home), especially if the damage is extensive. Not seeing the damage makes it harder for a builder to spot the cause of a moisture intrusion problem, which is why it's best to have your contractor involved from the beginning of the remediation process.

Some states (including our home state of Texas) require a licensed remediator to do the cleanup work and a separate contractor to handle the build back. This is to prevent dishonest companies from ripping out needless amounts of a home so that they can charge more for an extensive build back. In Texas, anything over twenty-five contiguous square feet warrants a professional remediation by a licensed, bonded, and certified company. These remediation experts should perform a mold air test before the remediation starts and after work is completed to make sure the moldy situation was resolved. Check local guidelines for more information about your unique situation.

STEP 5: FIND PROFESSIONAL CLEANING SERVICES

If you have a new leak and the mold hasn't been present for very long, you may be able to get away with washing your belongings in the washing machine on hot using detergent and some bleach to disinfect them. Follow with another wash cycle with detergent and one cup of clear ammonia. (The ammonia will help remove mycotoxin residues and buildup within the fabric's fibers.)

Depending on the extent of the damage, however, it may be wise to have your clothing, rugs, bedding, and upholstery fabrics cleaned by a professional remediation company using hydroxyl chambers.

Sometimes it's better (and more feasible) to simply throw away old upholstered items that are hard to clean, such as curtains, mattresses, and sofas. Even sheets and clothing can have wrinkles or sealed compartments that can absorb spores and mycotoxin residues.

Books are also hard to clean because paper is porous and very absorbent. Photos and other paper mementos may yellow around the edges because of mycotoxins. Have your precious family heirloom photos scanned and digitally stored so that you can throw away your old pictures and have new ones printed.

A 1:1 solution of water and clear ammonia in a spray bottle can be used to remove mycotoxins from solid surfaces that don't have added finishes (for example, glass, metal, or porcelain). Be sure to use clean paper towels and to spray the surfaces in a well-ventilated area, preferably outside.

GEEK BOX | hydroxyl

Hydroxyl molecules form when ultraviolet rays react with water vapor, and they survive just long enough to form hydrogen bonds and trigger the decomposition of harmful compounds. As such, this chain reaction kills bacteria, viruses, nasty odors, and more, including (you guessed it) mold! That said, hydroxyl molecules do not form naturally indoors, requiring special gear called hydroxyl generators, which are often used to ensure household items are free of mold.

STEP 6: SEEK MEDICAL HELP FROM A PRACTITIONER FAMILIAR WITH MOLD TOXICITY

A mold problem in a home will uproot a family and disrupt their daily lives. Jen was in her last semester of graduate business school—during final exams, no less—when her whole world unraveled. Once she finally found the toxic black mold in her apartment, she relocated to another apartment with her cats, her laptop, and a sack of clothes that were recently laundered at the local cleaners. (She eventually had to throw away many of the items that were professionally cleaned because the mycotoxins could not be completely removed.)

Jen's body had become riddled with the poisonous substances emitted by molds. Mycotoxin vapors penetrate building materials such as sheetrock and can attach to clothing, hair, and fabric. They even form films on hard surfaces such as metal, stone, wood, and glass. Humans and pets readily absorb these oily substances through the skin, eyes, lungs, and gastrointestinal tract—the body may even mistake mycotoxins for beneficial substances and recycle them from within. Depending on the potency of the mold strain and the specific types of chemicals the fungus released as airborne toxins, chronic exposure can lead to toxic levels in some individuals.

There is never a good time for this kind of toxic onslaught to happen. We all have responsibilities to ourselves, to our careers, and to those who depend upon us, but mold compromises the immune system and impairs the central nervous system, resulting in systemic inflammation, an inability to focus (brain fog), and extreme fatigue.

Coping with a catastrophic mold event that impairs one's health on a chronic level is one of the most trying things that can happen in a person's life. It affects not only you but also your loved ones, coworkers, and friends. Family members need love and support but often feel abandoned and confused when a loved one becomes ill. It's hard for others and even some doctors to understand this painful process, as the health problems caused by chronic indoor mold exposure can vary widely among individuals. Figuring out the problem can be frustrating and often ends in misdiagnosis. No matter the situation, we cannot stress enough the importance of not falling into a victim mentality, because it can create a downward spiral. This also means not letting your emotions get the best of your decision-making abilities.

Your best bet is to find a reputable environmental doctor who is familiar with both eastern and western medicine and understands that mold toxicity can impair the immune system and cause hypersensitive reactions. The limbic system goes on high alert, and your body gets stuck in the fight-or-flight mode. Environmental doctors and some functional practitioners are well-versed in diagnosing and treating mold toxicity. Remember that antibiotics and other western medicines are often necessary to treat chronic bacterial, fungal, or viral infections due to an impaired immune response.

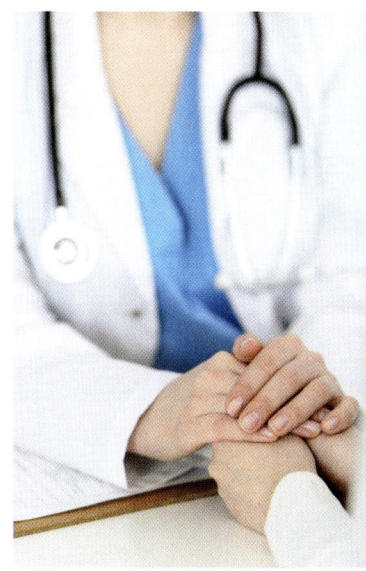

Other treatments to discuss with your doctor may focus on boosting and retraining the immune system. These can include gamma globulin infusions, T-cell harvesting, and vitamin IVs for those with leaky gut issues, which are common with chronic mold exposure.

Long-term immunotherapy (aka allergy shots) can be self-administered at home under a doctor's guidance. These shots can effectively neutralize allergic symptoms, which is amazing for anyone with chronic allergy issues, and they often work even better in combination with over-the-counter allergy medication so that the body can effectively detox and get back on track.

Select environmental health clinics run by well-trained doctors have state-of-the-art testing procedures and use allergy shots to effectively neutralize reactions to nearly everything in a person's environment that could be causing sensitization. These range from the traditional trees, grasses, weeds, and molds to foods, chemicals, mycotoxins, fabrics, supplements, and even sunrays (for those allergic to the sun!). It's important to note that neutralization allergy shots are designed to find the exact dose that eliminates symptoms from a specific irritant. These shots

GEEK BOX | leaky gut

Leaky gut occurs when ingested toxins create tiny perforations within a person's intestines. These holes are just big enough for partially digested food, toxins, and bacteria to get through and end up in the person's bloodstream. If this happens to you, your body will sound the alarm against the foreign invaders. The body produces a 24/7 immune response to these trespassers that can result in chronic inflammation, which will lead to a laundry list of medical issues. This is often how someone suddenly seems to become allergic to everything overnight.

are different from those that many allergy clinics provide, which feature a buildup phase and a maintenance phase and can exacerbate issues in a mold-sensitized person. Know which type of allergy shots you're getting.

Concentrated oxygen therapy can be administered via O_2 that is inhaled through a mask or inside a hyperbaric chamber. Oxygen is often integral to a functional medicine doctor's prescribed therapies. Chronic inflammation can cut off the delicate highway of vessels that make up a person's microcirculation system. All parts of the body need adequate blood flow to remain healthy, and forced oxygen therapy can help break the inflammation process.

Often, dormant Lyme pathogens or viruses from a previous infection can suddenly manifest symptoms again after mold exposure due to the body's impaired immune response. If all the immune cells are busy chasing down toxins and unwelcome contaminants, such as partially digested food proteins in the blood stream, then the body may begin to overlook the other bad guys it's kept in check. Ridding the body of chronic infections is an important step. Antibiotics and antifungals may be prescribed by your doctor, and ozone IVs are one of the therapies used to treat chronic vector type illnesses.

Another vital piece that is often overlooked by doctors is adequately monitoring thyroid function. The standard thyroid test may not pick up on disruptive hormone levels, so ask for a full thyroid panel that includes free T3 and free T4. These "frees" are the available thyroid hormones that your body can use. (Bound hormones are not usable but, oddly enough, are often the levels that are tested.)

And don't forget about your dentist. Your teeth and gums are a direct line to your bloodstream. Any kind of tooth infection or decay should be treated or removed promptly. Amalgam fillings (mercury mixed with silver) should be replaced by a dentist who specializes in these types of procedures, such as a biological dentist. Root canals are a no-no since they involve leaving a diseased or dead tooth in the gums and filling it with toxic resin and formaldehyde. A root canal tooth can also become a hotbed for bacterial infection, which will eventually wear away at the immune system. Pulling a tooth and using a biologically compatible replacement material is generally a cheaper and healthier solution to dental problems.

EATING BETTER FOR BETTER HEALTH

We also recommend eating a diet that is as close to 100 percent organic as possible. Even the very low levels of pesticide residues found on most conventional produce at the grocery store can add to your body's toxic load and create a roadblock to regaining health.

One factor that can help you heal is finding a nutritionist to guide you through the necessary lab tests and options for specific nutrients, which aren't always obvious. But not all nutritionists are trained in treating those with health issues caused by mold, and the process gets even more difficult when you consider trying to help someone who seems allergic to everything. A rotation diet (discussed in greater detail in the previous chapter) can help keep food allergies from getting worse and can even help them go away over time. If a hypersensitive person becomes sensitive to fabrics in clothing, sheets, or towels, you can also rotate the types of materials.

Of equal importance are the effects of accidentally ingesting heavy metals, which can lead to similar health issues (although even someone without a leaky gut can absorb these because the body often mistakes them for beneficial minerals—for example, lead mimics calcium). Existing heavy metal burdens within a person's body will complicate mold toxicity. Or, if the body already has difficulties absorbing nutrients due to a leaky gut caused by mold, heavy metals can enter and wreak havoc more easily. Bottom line: it's important for everyone to eat right and avoid pollutants, but it is especially important for people with mold toxicity.

THE BEST WAYS TO DETOX

If you're working to get rid of a mold problem in your living environment, chances are you're ready and willing to remove other problematic environmental toxins that may be pouring fuel on the fire. There's no point in doing these therapies and treatments unless you're taking steps to detox your body. A balanced therapeutic program to overcome mold toxicity should include a combination of detoxing factors.

Nontoxic, solid wood saunas that are mechanically joined (with no adhesives used) are excellent for long-term detox. Some people start once a week or once a month and work up from there. Dry saunas are the sweat boxes of choice and are very effective at "melting" and mobilizing toxins that get lodged in fatty tissues so that they can be released through sweat. Some toxins may still melt into the bloodstream, so you can also take a binder such as coconut charcoal capsules, bentonite clay, or psyllium husk fiber. Make sure your sauna manufacturer uses solid white aspen or poplar, two popular sauna woods that have become the gold standard for healthy infrared detox systems.

Finally, rest and relaxation are a real thing. Yoga, meditation, and massage are instrumental in restoring a sick body and maintaining health. Forcing the autonomic nervous system out of fight-or-flight and into a state of restful relaxation allows the body to repair itself. Getting enough restful sleep in this relaxed state is essential to maintaining healthy detox and cell rejuvenation.

Remember that restoring your health and life is possible! The goal with all these therapies is to desensitize, fortify, and detox. The human body is a remarkable machine with restorative abilities far beyond anything modern science can fathom. There are people who have used the same heart, lungs, and vascular system for over a century. We are designed to continually repair ourselves! A person who is dedicated to restoring their health—and who can maintain a positive outlook after a toxic mold insult—will overcome any illness related to mold toxicity.

Bear in mind that mold-sensitized people who have recovered often notice that they are more sensitive for the rest of their lives. Try to look at the positive side. Your body has a heightened sense of awareness to things that may pose a threat. Trust your instincts. Sure, allergic reactions can be annoying, but now you know when to leave a space before most others even notice something's wrong. You've become the canary in the coal mine—the barometer that measures whether your home is healthy. And maybe that's not such a terrible thing.

WHEN TO CONTACT A LEGAL PROFESSIONAL

If you have reason to believe that the mold infestation in your home or rental was due to negligence, then it may be time to contact a legal professional who is licensed in your state or jurisdiction. Remember that you must be able to prove there was negligence during the construction of your home or that your leasing management company did not follow protocol to remediate previous or active leaks that they were notified about in writing. Do your homework when choosing legal representation, and interview credible experts with experience in the toxic torts arena.

Also, think long and hard before jumping to legal action. The last thing most people want is to spend thousands of dollars to defend a case that has no burden of proof. Keep in mind that even the best builders and subcontractors cannot guarantee a home will be impervious to leaks or mold. There are simply too many factors outside of their control, including weather, ground shifting, and even homeowners who damage their own homes. That said, a contractor could have cut corners, installed something improperly, or knowingly overlooked a problematic situation, and in those cases a lawsuit might be warranted.

key points

▸ If you have an issue with mold, make plans to relocate and find an expert who can test your mold problem.

▸ Next, contact your insurance company for more information about coverage.

▸ Hire a reputable, licensed mold remediation company and a general contractor to do the repairs (we recommend hiring both at the same time).

▸ Finally, hire professional cleaning services and find a doctor who's familiar with mold toxicity.

You're almost done! Head to the epilogue to find out how to start your journey.

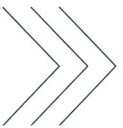

epilogue

starting your journey? here are the next steps

So, you're ready to build your new healthy home! Or perhaps you're already in the throes of design and construction.

We would be remiss if we didn't include a final piece about why this book is only the beginning of what a healthy home truly encompasses. It's more than shelter. More than a place to sleep. More than a place to relax and be creative. More than a lifetime of memories under one roof.

A healthier home is a way of life. And healthier living is a mindset.

As the founders of JS2 Partners, we saw the need for a change in the way homes are built. That need was originally born out of necessity for Jen's health, but the concept of cleaning up the insides of our houses was long overdue.

Construction and real estate are responsible for billions of dollars in sales each year. The construction industry has long been set in its ways when it comes to how materials are manufactured and how structures are designed and built.

Our team's focus from day one has been to create better living through healthy building.

The result has been larger than we ever imagined. A pivotal shift is happening in the construction industry. We can't take all the credit, but we aim to continue redefining and implementing home building best practices. And we thank our lucky stars that we're able to make an impact that is changing the way we all look at our indoor environments.

In short, "healthy home" is starting to become a buzzword. JS2 Partners has been contacted by homeowners in Texas (where we're headquartered) and almost every other state across the US about building healthy homes.

HOW HEALTHY HOMES CHANGE LIVES

We will never forget the cozy modern farmhouse ranch we built for a couple who thought moving back to Texas to be with their kids was impossible. The wife had experienced a decade-long battle with Lyme disease, allergies, and chemical sensitivities. She never thought she could live in a brand-new home and feel amazing.

We also recently sold a home to a young family from Michigan. They had lived in a moldy house that made Mom incredibly sick. Her body became too sensitive to inhabit an average home, as the chemical off-gassing from new construction would have caused her body to shut down. The family looked at many homes that were several years old, and each one had an underlying mold issue. They contacted us, and within two months they had moved to the Texas Hill Country to live in one of our new healthy homes. (This was the same house with the super low mold test results we mentioned earlier.)

But our homeowners aren't just people with allergies who are willing to move across the country. Our homeowners are wellness-conscious young families, empty nesters who want quality-built custom homes they know will last, and people who see their homes as an investment in their quality of life as much as a real estate asset.

In other words, our homeowners are just like you: unique! And just like you, they're already ahead of the curve when it comes to living a wellness-focused lifestyle.

Nonetheless, there are things you can do today to get into that healthier living mindset. Let's recap a few of the larger concepts—and discuss a few new ones as well.

THE BASICS OF BUILDING A HOME

Building a home is a complex undertaking that requires extensive planning from start to finish. Whichever step you're on, you can benefit from a quick review of each part of the process.

1.

When **breaking ground**, remember to perform a soil test and exercise caution to avoid underground obstacles and sources of radon. Get to know your property with professional land surveys, and place your home on the highest, flattest part to ensure proper drainage. Don't hesitate to hire a civil engineer to resolve any issues quickly and easily.

2.

Laying the **foundation** is one of the most important steps in building a house. We recommend slab-on-grade foundations, but if you require a pier-and-beam one, just be aware of the risks and make sure it's properly engineered. Avoid placing wiring or water mains inside the slab, and remember to install a radon mitigation system.

3.

Next comes **structure**, the bones of the home. Make sure you're familiar with local building codes and lumber availability before starting framing. (Avoid steel frames and alternative building materials like ICF, as they don't meet our healthy home standards.) Check your planks for mold, but don't bother with a tarp.

4.

Roofing and **exterior** sheathing are equally vital. This is the home's skin, protecting it from weather and other external forces. We prefer standing seam metal roofs, but make sure they're not flat or angled in ways that shed water poorly. For walls, ensure an effective vapor barrier is installed correctly, and always over-flash gaps in the armor such as windows and doors.

5.

Getting the **interior** right is the final step in construction. Always look for proper thermal breaks where insulation is required, and source natural drywall to avoid toxic chemicals. When it comes to the areas inside your home, it's just as important to use healthy materials as code-compliant building practices.

GOING BEYOND CONSTRUCTION MATERIALS

You know by now that it's essential to carefully consider what you bring into your home. We've talked about how to approach home building in the most healthful, quality-driven manner possible—but it doesn't stop there.

Furnishings, mattresses, rugs, curtains, cookware, and artwork are just as much a part of your everyday living space as the soap you use to clean the floor, the detergent you use to wash your clothes, and the foods you eat. Everything you bring into your home has a direct effect on your body and your quality of life.

We recognize that each home is a unique expression of its owners' personalities and a collection of their lives' contents. Being a part of the interior design process is just as much a part of our job as construction.

Little-known fact: we have been curating gorgeous, stylish, trendy, and timeless whole-home interior design packages for our clients for years. The composition of your home's interior furnishings is just as important as the paints, flooring, and countertops you select.

Here's a handful of questions to consider:

Is your wood furniture finished with a zero-VOC lacquer or preservative-free wood oil?

Are your rugs and curtains free of chemical fire retardants?

Is your sofa free of stain-resistant chemical treatments? (Upon skin contact, such treatments can transmit harmful chemicals.)

Are your couch cushions made of CertiPUR-US foam? (These foams are made without formaldehyde and other harmful chemicals.)

Are your pots and pans Teflon-free? (Teflon coatings can flake over time, getting into your food, and are linked to cancer.)

Are you using natural cleaners such as vinegar and hydrogen peroxide? It breaks our hearts to see a homeowner use toxic cleaning products in a brand-new healthy home.

Are your mattresses made of organic fibers or natural latex? (Most mass-manufactured mattresses off-gas PVC fumes and leach arsenic, which is applied as a fire retardant.)

Chances are your bookcases and end tables are made of MDF, particleboard, and/or engineered wood. If you build a healthy home with solid wood and nontoxic paint, shouldn't your furniture be the same?

Are your dishes and glassware lead-free?

Learn more about things to consider when furnishing your home in Part 5 of this book.

A GUIDE TO LABEL LINGO

When you are looking for clean and healthy products and home furnishings, labels can be challenging to understand—especially when you realize that many labels are, in fact, meaningless.

This is not only because US regulations fail to include many harmful VOCs on their lists of prohibited substances, but also because there are currently no official standards around many of the labeling *terms* used. In short, a lot of them are marketing and nothing more.

So, what should you be looking for? Here's a quick guide to a few of the most common phrases you might see on product labels in the US.

organic

The USDA governs whether a product can be labeled organic, which means free of synthetic materials such as antibiotics or pesticides. Unfortunately, studies have found that some organic products are mislabeled or cross-contaminated through shipping and handling processes. We source numerous organic materials when available.

green

If a product is "certified green," it has a reduced environmental impact, often through local sourcing and renewable or recyclable ingredients. But pay attention to who certified it, and keep in mind that green doesn't mean nontoxic.

natural

Usually, this label means a product claims to not contain anything artificial or synthetic, but the term is often ambiguous and misleading. Sometimes, a product with the "all natural" label simply isn't.

free of

Be wary of any label bragging about what's not in a product. Although helpful for allergen warnings, many products claiming to be free of a specific chemical contain alternatives that are just as harmful.

nontoxic

This term often means only that a product claims to be safer than some others and may not cause immediate adverse health effects—not that it is completely nontoxic, as there are no specific government standards for this term.

DOING IT YOURSELF

We're enthusiastic advocates for all you DIYers out there, and we admire you for your pride in getting your hands dirty and showing your passion in the form of sweat equity. DIY renovations are among the great traditions and rites of home ownership. However, a word to the wise: you should always exercise caution when tinkering with the structure in which you live—especially when upgrading an older house or a home that wasn't built with occupant health in mind.

For example, putting a new coat of zero-VOC paint on the walls is a simple DIY project, but if you need to cut into those walls for any reason, you risk releasing harmful chemicals or debris, from lead to old insulation—or, even worse, damaging the structural integrity of the home. For these reasons and for your own safety, we recommend reaching out to a professional builder or general contractor to do the specialized work.

NECESSARY MAINTENANCE

Finally, a word on wear and tear. Every home will deteriorate over time as the elements and the people living inside it take their toll. Here are a few tips for keeping your home as healthy as can be down the road:

- **Keep up with cleaning.** As we mentioned earlier, keeping a home clean limits the amount of mold spores, dust, and allergens that can get trapped inside. This includes having the HVAC units and air ducts cleaned from time to time by a professional.

- **Watch for moisture.** Maintain your roof, drainage systems, and interior sources of moisture to reduce the conditions that create mold. Do this through adequate waterproofing, temperature control, and ventilation.

- **Ensure plenty of ventilation.** Keep the air flowing through your home by using appropriate exhaust fans and opening windows.

- **Limit contaminants.** This one might seem obvious, but as we discussed earlier in this chapter, it's not always clear what we're bringing into our homes when we buy a new sofa or rug or try out a new cleaner. Every. Thing. Counts.

- **Create a maintenance schedule.** Your healthy home is an investment, and an ounce of prevention is always worth more than a pound of cure.

TAKING THE NEXT STEPS

This book is the beginning of your journey to a healthier home, and we are ready to embark on this journey with you even after you're done reading.

We love what we do, and we're passionate about continuing to pioneer the ways in which people live. This is exactly why the Healthier Homes online community was born. We wanted a way for you, the reader, to have access to up-to-date information and a place for our community of leaders and contributors to connect and share with confidence and support.

A part of our job as healthy home builders is to stay on top of new legislation, codes, and building standards that affect how homes are built. Often, it's our job to figure out the best way to pivot our models to meet healthy building standards while complying with new industry requirements. The world of home building and interior design is dynamic and ever-evolving, which is why we continue to add to our series of Healthier Homes eGuides, which are designed for readers who want to dig deeper into specific discussions and topics of interest. Knowledge is power, and we're all about empowering the next generation of homeowners.

We also recognize how hard it is to source high-performance, nontoxic building materials, which is why we created our own line of construction products. We use our Healthier Homes paints, primers, adhesives, cleaners, sealers, and more in the field every day. Transparency and quality are what we're all about. But perhaps the most exciting part of our new online hub is the shopping experience, where collections of items are curated by our top designers, and you can shop with peace of mind knowing that what you're getting will be a high-quality addition to your healthier home!

It's our dream to make all of this available online, all in one place, and we're continuing to add more of what you want, so please don't hesitate to reach out and tell us what you'd like to see.

At Healthier Homes, our mission is to provide a healthier living environment for you. See you at HealthierHomes.com.

Welcome home!
—Jen & Rusty

REFERENCES

PREFACE

1. Jack Wong Ho, "Fungal Toxins," *Handbook of Biologically Active Peptides,* ed. Abba J. Kastin (Elsevier Inc., 2013), 166–8.

2. Jeroen Douwes, Wijnand Eduard, and Peter S. Thorne, "Bioaerosols," *International Encyclopedia of Public Health,* ed. Stella R. Quah (Elsevier Inc., 2017), 210–8.

3. David William Austin et al., "Genetic Variation Associated with Hypersensitivity to Mercury," *Toxicology International* 21, no. 3 (2014): 236–41.

4. "MTHFR Gene, Folic Acid, and Preventing Neural Tube Defects," Centers for Disease Control and Prevention, page reviewed July 6, 2020, https://www.cdc.gov/ncbddd/folicacid/mthfr-gene-and-folic-acid.html

CHAPTER 1

1. "Constitution," World Health Organization, accessed March 18, 2022, https://www.who.int/about/governance/constitution

2. Joseph G. Allen et al., "36 Expert Tips to Make Your Home a Healthier Home," Harvard T. H. Chan School of Public Health, May 2019, https://forhealth.org/wp-content/uploads/2020/02/Harvard_Healthy_Buildings_Program_Homes_for_Health_May-2019_R1.8.pdf

3. "The top 10 causes of death," World Health Organization, December 9, 2020, https://www.who.int/news-room/fact-sheets/detail/the-top-10-causes-of-death

4. Rebecca Harrington, "The EPA Has Only Banned These 9 Chemicals—Out of Thousands," *Business Insider,* February 10, 2016, https://www.businessinsider.com/epa-only-restricts-9-chemicals-2016-2?op=1

5. "Toxicological Profile for Formaldehyde," Agency for Toxic Substances and Disease Registry, page reviewed May 12, 2015, https://wwwn.cdc.gov/TSP/ToxProfiles/ToxProfiles.aspx?id=220&tid=39

6. "Organic Solvents," National Institute for Occupational Safety and Health, Centers for Disease Control and Prevention, page reviewed November 2, 2018, https://www.cdc.gov/niosh/topics/organsolv/

7. "1,1,2-Trichloro-1,2,2-trifluoroethane," National Institute for Occupational Safety and Health, Centers for Disease Control and Prevention, page reviewed October 30, 2019, https://www.cdc.gov/niosh/npg/npgd0632.html

8. "Toxicological Profile for Total Petroleum Hydrocarbons," Agency for Toxic Substances and Disease Registry, page reviewed March 18, 2014, https://wwwn.cdc.gov/TSP/ToxProfiles/ToxProfiles.aspx?id=424&tid=75

9. Laura N. Vandenberg et al., "Low Dose Effects of Bisphenol A: An Integrated Review of In Vitro, Laboratory Animal, and Epidemiology Studies," *Endocrine Disruptors* 1, no. 1 (2013): e26490. 1–20.

10. Joseph G. Allen and John D. Macomber, Healthy Buildings (Harvard University Press, 2020): 127.

11. Antonia M. Calafat et al., "Polyfluoroalkyl Chemicals in the U.S. Population: Data from the National Health and Nutrition Examination Survey (NHANES) 2003–2004 and Comparisons with NHANES 1999–2000," *Environmental Health Perspectives* 115, no. 11 (2007): 1596–602.

12. Gang Liu et al., "Perfluoroalkyl Substances and Changes in Body Weight and Resting Metabolic Rate in Response to Weight Loss Diets: A Prospective Study," *PLoS Medicine* 15, no.2 (2018): e1002502.

13. Philippe Grandjean et al., "Serum Vaccine Antibody Concentrations in Children Exposed to Perflourinated Compounds," *Journal of the American Medical Association* 307, no. 4 (2012): 391–7.

14. Craig Butt and Heather Stapleton, "Inhibition of Thyroid Hormone Sulfotransferase Activity by Brominated Flame Retardants and Halogenated Phenolics," *Chemical Research in Toxicology* 26, no. 11 (2013): 1692–702.

15. Julie Herbstman et al., "Prenatal Exposure to PBDEs and Neurodevelopment," *Environmental Health Perspectives* 118, no. 5 (2010): 712–9.

16. John Meeker et al., "Polybrominated Diphenyl Ether (PBDE) Concentrations in House Dust Are Related to Hormone Levels in Men," *Science of the Total Environment* 407, no. 10 (2009): 3425–9.

17. Courtney Carignan et al., "Urinary Concentrations of Organophosphate Flame Retardant Metabolites and Pregnancy Outcomes among Women Undergoing in Vitro Fertilization," *Environmental Health Perspectives* 125, no. 8 (2017): 119001. PMID: 28858831; PMCID: PMC5783651.

18. "Public Safety and Homeland Security," Federal Communications Commission, accessed March 18, 2020, https://www.fcc.gov/public-safety-and-homeland-security

19. Wei-Jia Zhi, Li-Feng Wang, and Xiang-Jun Hu, "Recent Advances in the Effects of Microwave Radiation on Brains," *Military Medical Research* 4, no. 29 (2017). https://doi.org/10.1186/s40779-017-0139-0

20. Marjo Nikulin et al., "Stachybotrys atra Growth and Toxin Production in Some Building Materials and Fodder under Different Relative Humidities," *Applied and Environmental Microbiology* 60, no. 9 (1994): 3421–4.

21. Marina Leino et al., "Intranasal Exposure to Stachybotrys chartarum Enhances Airway Inflammation in Allergic Mice," *American Journal of Respiratory and Critical Care Medicine* 173, no. 5 (2005): https://doi.org/10.1164/rccm.200503-466OC

CHAPTER 3

1. "Altitude as a Factor in Air Pollution," United States Environmental Protection Agency, accessed March 21, 2022, https://cfpub.epa.gov/ncea/risk/recordisplay.cfm?deid=47804

2. Sivani Saravanamuttu and Sudarsanam Dorairaj, "A Survey on the Impact of Radiofrequency Electromagnetic Fields (RF-EMF) from Wireless Devices on Information Technology (IT) Professionals," *European Journal of Experimental Biology* 6, no. 4 (2016): 46–51.

3. Olle Johansson, "Electrohypersensitivity: State-of-the-Art of a Functional Impairment," *Electromagnetic Biology and Medicine* 25, no. 4 (2006): 245–58.

CHAPTER 5

1. Joseph G. Allen and John D. Macomber, *Healthy Buildings* (Harvard University Press, 2020): 51.

2. Shu-Ye Jiang, Ali Ma, and Srinivasan Ramachandran, "Negative Air Ions and Their Effects on Human Health and Air Quality Improvement," *International Journal of Molecular Sciences* 19, no. 10 (2018): 2966.

CHAPTER 9

1. Joseph Lstiburek, "BSD-106: Understanding Vapor Barriers," Building Science Corporation, April 15, 2011, https://www.buildingscience.com/documents/digests/bsd-106-understanding-vapor-barriers

2. Joseph G. Allen et al., "The 9 Foundations of a Healthy Building," Harvard T. H. Chan School of Public Health, 2017, https://forhealth.org/9_Foundations_of_a_Healthy_Building.February_2017.pdf

CHAPTER 13

1. Osman Can Ozcanli, "Turning Body Heat into Electricity," *Forbes,* June 8, 2010, https://www.forbes.com/2010/06/07/nanotech-body-heat-technology-breakthroughs-devices.html?sh=4350e27f387c

2. "Electromagnetic Fields and Public Health: Mobile Phones," World Health Organization, October 8, 2014, https://www.who.int/news-room/fact-sheets/detail/electromagnetic-fields-and-public-health-mobile-phones

3. Rafał Pawlak, Piotr Krawiec, and Jerzy Żurek, "On Measuring Electromagnetic Fields in 5G Technology," Institute of Electrical and Electronics Engineers, *IEEE Access* 7 (2019): 29826–35.

CHAPTER 14

1. Filipa Gonçalves et al., "Evaluation of Antimicrobial Properties of Cork," *FEMS Microbiology Letters* 363, no. 3 (2016): fnv231.

2. Rune Becher et al., "Do Carpets Impair Indoor Air Quality and Cause Adverse Health Outcomes: A Review," *International Journal of Environmental Research and Public Health* 15, no. 2 (2018): 184.

3. "Hazardous Substance Fact Sheet," New Jersey Department of Public Health, accessed March 21, 2022, https://nj.gov/health/eoh/rtkweb/documents/fs/1786.pdf

CHAPTER 15

1. Brady Seals and Andee Krasner, "Gas Stoves: Health and Air Quality Impacts and Solutions," RMI, 2020, https://rmi.org/insight/gas-stoves-pollution-health/

2. Anne Mulkern, "California Is Closing the Door to Gas in New Homes," *E&E News,* January 4, 2021, https://www.scientificamerican.com/article/california-is-closing-the-door-to-gas-in-new-homes/

CHAPTER 17

1. James Tarbox, "The Role of Cotton in Respiratory Symptoms in the Fall," *Pulmonary Chronicles* (2017): https://pulmonarychronicles.com/index.php/pulmonarychronicles/article/view/422/937

2. Lauren Rosenthal and Will Craft, "Buried Lead: How the EPA Has Left Americans Exposed to Lead in Drinking Water," *APM Reports,* May 4, 2020, https://www.apmreports.org/story/2020/05/04/epa-lead-pipes-drinking-water#:~:text=Congress%20banned%20the%20use%20of,known%20as%20%22service%20lines.%22

CHAPTER 19

1. "2011 NSF International Household Germ Study," The Public Health and Safety Organization, 2011, https://d2evkimvhatqav.cloudfront.net/documents/2011_NSF_Household_Germ_Study_exec-summary.pdf

2. "Food Allergies in Children," Johns Hopkins Medicine, accessed March 24, 2022, https://www.hopkinsmedicine.org/health/conditions-and-diseases/food-allergies-in-children

INDEX